ROBOTS

The MIT Press Essential Knowledge Series

ROBOTS

JOHN JORDAN

The MIT Press | Cambridge, Massachusetts | London, England

This book was set in Chaparral and DIN by Toppan Best-set Premedia Limited. Printed and bound in the United States of America.

Library of Congress Cataloging-in-Publication Data

Names: Jordan, John M. author.
Title: Robots / John Jordan.
Description: Cambridge, MA : MIT Press, [2016] | Series: The MIT Press essential knowledge series | Includes bibliographical references and index.
Identifiers: LCCN 2016008323 | ISBN 9780262529501 (pbk. : alk. paper)
Subjects: LCSH: Robotics—Popular works. | Robots—Social aspects—Popular works.
Classification: LCC TJ211.15 .J67 2016 | DDC 629.8/92—dc23 LC record available at https://lccn.loc.gov/2016008323

10   9   8   7   6   5   4   3   2   1

CONTENTS

SERIES FOREWORD

The MIT Press Essential Knowledge series offers accessible, concise, beautifully produced pocket-size books on topics of current interest. Written by leading thinkers, the books in this series deliver expert overviews of subjects that range from the cultural and the historical to the scientific and the technical.

In today's era of instant information gratification, we have ready access to opinions, rationalizations, and superficial descriptions. Much harder to come by is the foundational knowledge that informs a principled understanding of the world. Essential Knowledge books fill that need. Synthesizing specialized subject matter for nonspecialists and engaging critical topics through fundamentals, each of these compact volumes offers readers a point of access to complex ideas.

Bruce Tidor
Professor of Biological Engineering and Computer Science
Massachusetts Institute of Technology

Attempting to write a book about robots represents a leap of faith. The field is already broad and expanding further, and it moves too fast for a multiyear project to be at all current. Why, then, did I proceed?

The field of robotics is, I believe, entering a crucial stage. Technologies are good enough for mass commercialization, use by governments, and even to become invisible in their ubiquity. Robots will soon affect millions of people's lives more directly and profoundly.

The technical choices made in robots' design not only embody value judgments and aspirations; they often have ethical implications. Every roboticist I have met is smart, humane, and well spoken. Even so, I do not want a small population of scientists and engineers, working in isolation, to make all the decisions that can affect life, death, health, work and livelihood, class status, personal privacy, gender identity, the future of war, the shape of urban landscapes, and many other domains: they need help, and other perspectives.

This book seeks to widen the circle of individuals who have a say in what robots can and should do, look like, include, and leave out. I hope roboticists read the book, but the intended audience is the rest of us. Design choices made now may well be with us for decades, so now is the time to ask, and to assert, what "good" robots might look like. Because of the breadth and dynamism of the field, my narrative's emphasis falls less on completeness and the latest developments, and more on the enduring issues: what are the persistent capabilities, contests, and trade-offs that will characterize robots and robotics?

Why do these issues matter? Many powerful scenarios involve humans and robots working in partnership rather than robots *replacing* humans, whether on a battlefield, in a hospital, on an assembly line, or in the rehabilitation, prosthetics, or aging processes. Rather than focusing on either-or debates over what constitutes a robot, we will be better served by seeing compu-mechanical augmentation of human traits on a continuum. By necessity, this implies both that robots and human beings will be living and working closely, changing the nature of the human condition in important ways, and that, rather than becoming only slaves or potentially overlords, robots will become humanity's partners. These impending changes make improving our theories, norms, and aspirations an urgent matter. This book, this leap of faith, is a small step in that direction.

ACKNOWLEDGMENTS

This slim volume, five years in the making, was only possible with the help of many people, not all of whom can be named here.

Katherine A. Almeida, Kate Hensley, and especially Marie Lufkin Lee at the MIT Press have been consummate professionals, helping me, by turns, with advice, encouragement, constructive challenges, and superb execution. I count myself fortunate to have had the backing of such a capable team.

When Bob Bauer showed me the personal robot PR2 at Willow Garage back in 2011, that was the precise moment—on a par with my first glimpse of the National Center for Supercomputing Applications (NCSA) Mosaic web browser—that I saw everything differently. Bob later introduced me to my key research interviewees, including Steve Cousins, Scott Hassan, James Kuffner, and Leila Takayama; without him, this book would never have happened, so my thanks are profound.

In addition to many anonymous readers from MIT Press who each improved this book, I can thank four readers by name. Steve Sawyer offered a generous, incisive critique of an earlier draft and dazzled me with the breadth of his suggestions. Kate Hoffman deepened my understanding of anime and science fiction, drawing on her vast trove of reading recollections and judgments. My longtime partner in tech analysis, John Parkinson, read multiple drafts and unfailingly helped me, whether by cracking macro framing problems or correcting imprecision in the details.

Finally, my onetime coauthor, David Hall, offered encouragement, critical reaction, and expert perspective every time I asked him, even though he had dozens of other irons in the fire. Death took him, suddenly and too soon, a month before this book went into production. I can only hope he would have been proud of the final result.

INTRODUCTION

Television and cinema have helped propagate powerful American stories of technology. The *Star Wars* franchise is in many ways an updated Western, using space as a frontier (though maybe not the final one) and light sabers as rifles. The impact of such icons—and possibly archetypes—as Robocop, *Blade Runner*'s replicants, the brilliant, mannerly C-3PO, and Disney's Wall-E is wide and deep. Ask about "throwbots," Atlas, Motoman, Kiva, Beam, or other real-life robots that are reinventing warfare, the industrial workplace, and human-robot collaboration, and most people have little sense of what real robots do or look like. But everyone knows the Terminator, complete with Austrian accent.

In 2004, Chris Van Allsburg's beloved children's book *The Polar Express* was made into a movie. Oscar-winning talent (including Tom Hanks) was wired up for digital motion capture, but the resulting characters came out, to use critics' words, "creepy," "eerie," and "dead-eyed"; the movie was "a zombie train." For years, digital animators pushed for more polygons, more shades of color, more pixels—in short, more computing. But instead of delighting audiences with verisimilitude, they entered the "uncanny valley," a paradox in which added artificial likeness weirds people out. After 2005, the Japanese female android Repliee generated the same reaction with its almost-but-not-quite humanlike features.

In contrast to Hollywood, Boston Dynamics (between 2013 and 2016, a Google company) builds robots intended for use by the U.S. armed services. Nonclassified YouTube videos of a robotic cheetah, humanoid, and pack animal have received tens of millions of views, giving many people their first look at state-of-the-art robotic science. Rather than the moderately large viewership, most striking to me was my students' reaction when a human shoved/kicked the BigDog to show its stability: they gasped, as if someone had struck a pet on camera.

Robots are becoming more numerous, more capable, and more diverse. Over the long term, their economic, civic, and destructive impact will likely be on a par with that of the automobile. In such a massive transition, people will care what happens and call for rules, norms, and paths of recourse. Citizens have a vested interest in work, wages, and workplace safety; in aging with dignity; in major changes in global warfare; in privacy; and in other things that robotics has the potential to change. Many factors, however, combine to make it hard to advance the discussions about what we want from today's and tomorrow's robots.

Barriers to Informed Dialogue

When an innovation emerges, the history of its naming shows how it goes from foreign entity to novelty to invisible ubiquity. A little more than 100 years ago, automobiles were called "horseless carriages," defined by what they were not rather than what they were. More recently, the U.S. military referred to drones as "UAVs," unmanned aerial vehicles, continuing the trend of definition by negation.

The word "robot" originated in the 1920s and was at first a type of slave; robots are often characterized by their capabilities in performing dull, dirty, or dangerous tasks, sparing humans the need to perform them. The science and engineering of the field continue to evolve rapidly—look no further than Google's self-driving car or the humanoid robots it acquired in the Schaft and Boston Dynamics deals. Given such rapid change, computer scientists cannot come to anything resembling consensus on what constitutes a robot. Some argue that a given device qualifies if it can (1) sense its surroundings; (2) perform logical reasoning with various inputs; and (3) act upon the physical environment. Others insist a robot must move in physical space (disqualifying the Nest thermostat), while still others say that true robots are autonomous (excluding factory assembly tools). *Reason 1 why robots are hard to talk about: the definitions are unsettled, even among those most expert in the field.*

Bernard Roth, a longtime professor of mechanical engineering who was associated with the Stanford Artificial Intelligence Laboratory (SAIL) from its inception, offers a more nuanced definition that draws on his long history in the field. Roth begins by doubting that "a definition [of what is or is not a robot] will ever be universally agreed upon." Instead, he argues for a much more relative and conditional approach: "My view is that the notion of a robot has to do with which activities are, at a given time, associated with people and which are associated with machines." When relative capabilities evolve, so do conceptions: "If a machine suddenly becomes able to do what we normally associate with people, the machine can be upgraded in classification and classified as a robot. After a while, people get used to the activity being done by machines, and the devices get downgraded from 'robot' to 'machine.'"[1] *Reason 2: definitions evolve, unevenly and jerkily, over time as social context and technical capabilities change.*

Expectations for robotics are different from those for every other new technology because the vocabulary of robotics is so deeply a legacy of science fiction, both in literature and movies and on television. No technology has ever been so widely described and explored before its commercial introduction: the Internet, cell phones, refrigeration and air-conditioning, elevators, atomic energy, and countless other innovations that remade lives and landscapes saw the daylight of commercialization relatively quietly. Only later, if at all, did fiction invent vast numbers of fantasies, reaching audiences in the hundreds of millions, with the technologies at the center. In sharp contrast, fiction preceded, and conditioned, the science and engineering of robotics in an unprecedented fashion. *Reason 3: science fiction set the boundaries of the conceptual playing field before the engineers did.*

This paradox relates in part to a historical accident: the means by which mass-market sci-fi was disseminated between 1940 and 2000—paperback and comic books, cinema, television—all came of age in that same period. Thus the technologies of mass media helped create public conceptions of and expectations for a whole body of compu-mechanical innovation that *had not happened yet*: complex, pervasive attitudes and expectations predated the invention of viable products.

Given that modern Western robotics has been so strongly influenced by science fiction, why might that

The technologies of mass media helped create public conceptions of and expectations for a whole body of compu-mechanical innovation *that had not happened yet.*

matter? A whole system of meanings and expectations has been created by fantasy rather than fact. Most important, science fiction has set expectations for "real" robots unrealistically high: nontechnical journalists, fiction writers, and even the Oxford philosopher Nick Bostrom ask if a robot might volitionally "turn" against its maker even when that would be technically impossible. The point here is that other cultural cargo was smuggled in with the stories and movies: assumptions about robots relative to work, warfare, and human ability or disability will need to be consciously revisited at the same time that we examine ethics, autonomy, and the supposed capacity for evil.

A further naming issue must be raised here. Although it has neither the fictional nor, especially, the cinematic legacy of the robots, the related notion of artificial intelligence creates instant confusion and distrust among many nonroboticists. Elon Musk, the CEO of SpaceX and Tesla Motors, said in a 2014 MIT symposium that AI might be humanity's "biggest existential threat":

> I think we should be very careful about artificial intelligence. If I were to guess like what our biggest existential threat is, it's probably that. So we need to be very careful with the artificial intelligence. Increasingly scientists think there should be some regulatory oversight maybe at the national and international level, just to make sure that we don't

do something very foolish. With artificial intelligence we are summoning the demon. In all those stories where there's the guy with the pentagram and the holy water, it's like, yeah, he's sure he can control the demon. Didn't work out.[2]

Note Musk's recourse to myth and fiction as he attempts to explain his perspective on a vast scientific landscape. Moving away from wizards and demons, recall that the biggest successes in the field of artificial intelligence have come in controlled, limited domains: chess, Go, and the *Jeopardy!* game show most visibly, but also type-ahead search fields and automated mobile ad placement. It is critically important to distinguish between artificial *general* intelligence (and eventually humanlike cognition) and the *domain-specific* algorithms developed to do tasks such as credit scoring, fraud detection, or Google map route planning, each of which is essentially worthless outside its tuned specialty.[3] That said, deep learning is getting a lot better very fast, but, as I will argue, maybe the human brain isn't the proper benchmark for success.

Yet the fear of artificial life-forms surpassing humans in capability is widespread. This fear persists even though extremely basic tasks—such as getting the printer to connect—remain challenging. Whether robotic technologies are called "AI," "robots," or superdeveloped "intelligent personal assistants," such as Apple's Siri or Spike Jonze's

Samantha in the film *Her*, matters only slightly. The fear and uncertainty generated by fictional representations far exceed human reactions to real robots, which are often reported to be "underwhelming."[4] *Reason 4: robotics bleeds into other technologies that are even more poorly understood and ominously portrayed in many cultures.*

Path Dependence

Every new technology that begins in a laboratory and enters wide use passes through a series of stages. Early on, first basic, then applied science can be prohibitively difficult. Engineering at this stage of development is hard, and the primary question at hand is "How can we make this work?" Slightly later, before entrepreneurs and others solve the business model question ("How might this technology make money?"), there comes a time when design decisions are made that will shape much of the technology's future impact. Alternating versus direct current electricity, railroad gauges, and typewriter keyboard layouts are all examples of an economic notion called "path dependence": today's choices are constrained by technical decisions made in the past.[5]

In the fields of artificial intelligence, robotics, sensors, and information collection/processing, among others, we are at a point where people beyond the engineers and scientists should be involved in the discussion. "How do we make it work?" is still a live issue, but we also have a choice among multiple paths forward. In short, it is time to start asking more people what they desire—and what they resist—from these technologies. Robotics has many potential implications: for economic livelihoods and wealth accumulation, identity and relationships, citizenship and warfare, privacy and individual agency. Design choices involve far more than only engineering constraints. Politics, economics, luck, and other forces are also at work but, thus far, are being addressed only peripherally.[6]

To make this notion less abstract, consider two examples. Jaron Lanier tells a story in his book *You Are Not a Gadget* that applies beautifully. When the musical instrument digital interface (MIDI) first connected synthesizers to computers, a design decision was made based on the state of computer science to make keyboard triggers essentially binary: either the key was "pressed" (in the digital domain) or it was not. Real music like the blues, however, allows musicians to "bend" or otherwise shape notes, such as on a guitar or harmonica. MIDI music cannot sound this way because of the path dependence related to the initial specification. As a result, we have had thirty years of what Lanier calls "beepy"-sounding electronic music, and it didn't have to be that way.[7]

More recently, Google tried to expand and integrate its social services such as YouTube, Gmail, and

GrandCentral (since 2009, Google Voice) into a network called "Google+," launched in 2011. For a number of reasons, the service insisted that people sign up with their real-world names and gender identity. Real-name identification had the upside of reducing flame wars in some online forums, and made life easier for Google to track user behavior across its properties for advertising purposes, but it posed a privacy nightmare for some people. At least one transgender person's sexual identity was revealed in a text message without consent because of Google's integration of Google+ information into Android address books. In 2014, Google cofounder Sergey Brin, who was intimately involved in Google+ design decisions, admitted, "I'm probably the worst person to speak about social ... I'm just not a very social person."[8] As with countless design choices, that decision had far-reaching consequences. In addition to personal inconvenience and worse, Brin's rigid insistence on real-name integration across Google's many websites alienated many users and probably contributed to Google+'s weak public reception. Later in 2014, Google abandoned the policy.

Why Robotics Matters: Some Concrete Considerations

How does robotics differ from what we came to know of digital computing from 1950 until 2005 or so? Here are just a few examples of big, complex issues we need to confront—and quickly.

1.   With cameras and sensors everywhere—on telephone poles, on people's faces, in people's pockets, in the ground (on water mains or monitoring seismic activity), and in the sky (drone photography has given rise to a rapidly evolving body of legal judgment and contestation)—the boundaries of privacy, security, and risk are all being reset. What are the rights of the observed and the responsibilities of the observer? (A powerfully dystopian vision of pervasive monitoring, peer pressure in social networks, and winner-take-all markets can be found in Dave Eggers's novel *The Circle*.)

2.   When robots enter combat, how and when will they be hacked? Who will program a self-driving suicide bomb? (The Islamic State was reported to be doing so in early 2016.) Are drone pilots or robot software writers subject to the Geneva Convention? When robots administer torture, who is responsible? This technology, so aggressively developed by the military, will open multiple debates in the realm of warfare and conflict.

3.   Computer science, information theory, statistics, and physics (in terms of magnetic media) are all being stress tested by the huge volumes of data generated by our

increasingly instrumented planet. Sensors are intrinsic to robotics; the two domains are often difficult to distinguish. A General Electric jet engine is reported to take off, on average, every two seconds, worldwide. Each engine generates a terabyte of data per flight.[9] Even when a 10 to 1 compression takes this figure down to 100 gigabytes per flight, that's about a million DVDs' worth of raw data every day. Because it cannot all be kept for any length of time, sampling, (further) compressing, logging, and other data actions must be perfected. Dealing with information problems at this scale, in nearly every domain, raises grand challenge–scale hurdles in business, academe, medicine, and even sports.

4. The half-life of technical knowledge appears to diminish ever faster. Machine learning, machine vision in robots, and other fields are evolving rapidly, complicating employment patterns and career evolution. Robots will obviously displace manual laborers, but engineers, programmers, and scientists will also be hard pressed to maintain current skills. In addition, we are increasingly dealing with platforms (such as Microsoft Windows, Apple iOS, and Google maps) rather than products. Platforms are far more powerful: as Chunka Mui and Paul Carroll point out, every Google self-driving car learns from the experiences of every other Google car.[10] Accordingly, how will we learn

to think about a world in which Google could put cars, robots, thermostats, watches, and phones on a common Android software base? Platform economics, involving as they do such key concepts as lock-in and license exclusivity, have powerful and far-reaching social effects.

5. What are the rules of engagement with computing that moves about in the wild? A woman wearing a Google Glass headset was assaulted in a bar after she violated an implied social contract; self-driving cars don't yet have clear liability laws; 3-D printing of guns and of patented or copyrighted material has yet to be sorted out; nobody yet knows what happens when people can invoke facial recognition of a stranger on the sidewalk; Google could see consumer (or EU) blowback if it uses Nest sensor data to drive ad targeting. As a reminder of how much these rules matter, when the telephone was first introduced, it was socially impolite to greet someone to whom you had not been introduced. As a result, many languages have two words of greeting, one for the phone (the French "allô") and one in person ("bonjour"). Alexander Graham Bell had his own preferred solution in English: using "ahoy" as the telephonic greeting.[11] One hundred thirty years later, we have much to negotiate in the next wave of physical computing, including the norms that apply to social life.

6. How will these technologies augment and amplify human capability? Whether through exoskeletons, care

robots, telepresence, or prostheses, the human condition will change in shape, reach, and scope in the next 100 years. At the same time, how will humans enable the new computing models? An ATM or self-driving car needs to be instructed, as does the da Vinci Surgical System (which, though robotic, does not really qualify as a robot). The potential for collaborations between human and compu-mechanical systems is mind boggling: how many more Stephen Hawkings, Adrianne Haslet-Davises,[12] or Robin Millars[13] will we see? But first, a variety of technical and nontechnical challenges must be identified, named, and negotiated (not just "solved").[14] And how will access to these augmentations be allocated?

7. Compared to keyboards, screens, and mice, robotics helps introduce a myriad of new ways for humans and machines to interact. Nods, winks, finger swipes, spoken commands, and even brain waves can generate actions. Given human variation, cultural differences, vast differences in languages, and physical constraints like power consumption or water resistance, how will humans learn to "drive" all these new tools? One fascinating set of decisions relates to color. Many mobile robots currently share a clinical white color, including Sony Aibo, Honda Asimo, Bestic, Jibo, Beam, and Atlas II. Recall that, for years, desktop PCs were uniformly beige, then black or gray became common until Apple redefined the color spectrum

with shades of turquoise and tangerine. Given that computers close to humans convey layers of meaning, it will be worth watching how the red Baxter robot is received, and if black or brown variations of autonomous robots might be developed, potentially for less Caucasian markets.

8. The infrastructure enabled—and required—by robotics and related fields (such as the Internet of Things or IoT) looks very different from that of an industrial economy. Systems get bigger as needs increase and technologies of management and control get better. In part, these changes reflect new risks. Robotic technologies also require different kinds of workplaces: no more need for warehouses to be air-conditioned for human comfort, but assembly-line bots need safety cages. Transit systems will change as robots on the move—whether on a roadway, in airspace, or in a hospital corridor—require different signals and safety precautions than human drivers do.

Together, these eight sets of questions involve law, belief, economics, education, public commons, public safety, and human identity along with technical fields of power management, magnetic storage, materials science, algorithmic calculation, and more. Given the breadth and magnitude of its impacts, robotics is simply too important, and now too close at hand, to be left only to technical specialists.

Summary

The laws, stories, economic forces, and blind spots accompanying robots and robotics are neither inevitable nor obvious. They take work to craft, untangle, and assess. The next wave of computing will introduce profound changes, which will eventually rival those brought about by the automobile, household electricity, or running water. (As an example, consider the profound impact of drone warfare in the absence of public or even congressional debate over its ethical, political, and strategic implications.) Given that the technologies for self-driving cars, embedded or face-mounted computers and sensors, and autonomous robots are all coming to market in a matter of months rather than decades, the circle of people involved in overseeing them needs to be widened. Time and again, engineers and scientists have answered the question, "How can we make this work?" Now it's time for more of us to ask, "What realistic choices can be made in each of these domains?"—and to help make those choices.

This process is anything but straightforward: people often don't know what they want when presented with known alternatives, as the study of "choice architectures" reveals.[15] As ways of assessing "new new" products, focus groups and other techniques for market research are fatally flawed. Prior to the launch of the iPhone, smartphone sales were largely confined to BlackBerry and Nokia devices, none of which had glass interfaces. Five years later, both companies had become essentially irrelevant as Apple and Google's Android mobile operating systems dominated the market even though both Research in Motion and Nokia spent heavily on research and development. So far, robot markets have yet to declare their preferences. That said, for private drones, face-mounted computers, and other technologies, the time has come to consider "rules of the road," outright prohibitions, and warning signs to be watched in the early phases of adoption.

The field of "robot ethics" addresses some of these issues.[16] One line of thinking dates back to Isaac Asimov, and considerable effort is still spent in wrestling with problems of human versus compu-mechanical agency, particularly with regard to human harm. People also have a deep-rooted habit of attributing cognizance and motive ("The machine is thinking") where neither is present. Closer to our own time, the moral dimension of military robots programmed to shoot first in the interest of self-preservation has drawn worldwide attention, as from the NGO group Campaign to Stop Killer Robots.[17] Elsewhere, Steven Pinker, a leading scientist of the brain, is clear in his stance on human moral responsibility: "Why give a robot an order to obey orders—why aren't the original orders enough? Why command a robot not to do harm—wouldn't it be easier never to command it to do harm in the first place?"[18] In addition, the emerging capabilities of artificial intelligence, a field with both contested self-definitions and broad application to robotics, make boundaries between humans and machines even fuzzier. Some of the design decisions that will be made in the coming years have high stakes, with consequences that involve the very essence of human agency, identity, and belief.

Robotics introduces a new layer of complexity into the fields of artificial intelligence, "big data," and ultimately, human meaning. Not only does the long history of human efforts to create artificial life see a new chapter, but also people can now create artificial life in vast networks that will behave differently than a single entity: Frankenstein's creature can be understood as a forerunner of the Boston Dynamics Atlas robot, but there isn't as visible a precedent for self-tuning sensor nets or swarms of self-organizing drones. The change is happening at a faster rate of technological innovation than humans are accustomed to, raising further complications.

This transition is in the early stages, and human intellect is often held up as the ideal to which machine designers should aspire. Ray Kurzweil's entire singularity notion is premised on "the idea that we have the ability to understand our own intelligence—to access our own source code, if you will—and then revise and expand it."[19] At the same time, machines that surpass humans in capability represent a new chapter in the "technics out of control" meme that the historian of technology Langdon Winner has so ably cataloged. Although caution in the early adoption of any new technology is prudent, as examples from nuclear power clearly show, our fear of robotic technologies is largely misplaced. According to Rodney Brooks, the CEO of Rethink Robotics and a longtime MIT researcher in the field, achieving AI is really difficult. For silicon circuitry to develop conscious malevolence is likely at least a century off. As Brooks noted in November 2014, "If we are spectacularly lucky we'll have AI over the next thirty years with the intentionality of a lizard, and robots using that AI will be useful tools. And they probably won't really be aware of us in any serious way. Worrying about AI that will be intentionally evil to us is pure fear mongering. And an immense waste of time."[20]

At some levels, humanity is at a stage somewhere between da Vinci and the Wright Brothers in the development of powered flight. Airplanes do not fly by flapping wings, nor do birds carry 500 people 9,500 miles. Projects to reverse engineer the human brain, running as it does more on chemistry than on electrons, seem to be limited in their applicability. Instead of Hollywood archetypes and linguistic allusions, we need the lab work for which Orville and Wilbur Wright are less appreciated. They invented not just the airplane, but significant advances in the science of aeronautics (as well as the method for flying the airplane.) The human metaphors embedded in robotics and artificial intelligence are shaping progress in this field, likely holding us back as much as they inspire. Twenty-second-century AI will probably no more mimic the human brain than the airplane mimics a bird, or a wheel mimics a leg. Abstracting the problem beyond biomimicry represents an initial step in this process: Who will build the equivalent of the wind tunnel for cognition?

Regardless of the state of computing, science fiction, or filmmaking, we are about to colonize new territory on the technological frontier. And though it is appropriate to honor the pioneers, it is equally appropriate for the settlers to have a voice in the laws, customs, economics, and social conventions that we will observe. Because we live in ever closer proximity to computing that inhabits and transforms our physical world, it is time to question the mental models we use to describe robotics.

THE PREHISTORY OF AN IDEA

Humans have attempted to re-create life for millennia, and many robots of the early twenty-first century continue these traditions. The context for today's efforts is important to acknowledge, especially given the both problematic and enduring character of some of these discussions: the impact of Mary Shelley's *Frankenstein* is difficult to overstate, to take only one example.

Words

Before we look at history, a bit about terminology: for such a familiar concept, it turns out that robots are extremely difficult to define. According to the *American Heritage Dictionary of the English Language* (3rd edition), a robot is "a mechanical device that *sometimes resembles a human being* and is capable of performing a variety of often complex human tasks on command or by being programmed in advance" (emphasis mine). This biomimicry raises one set of issues, especially in regard to autonomous robots. The *Oxford English Dictionary* definition introduces a second, literary area of complexity: "Chiefly *Science Fiction*. An intelligent artificial being typically made of metal and resembling in some way a human or other animal."

Roboticists themselves struggle to pin down the definition of their field. George Bekey, an expert in autonomous robots, defined a robot by its characteristics: sensing, artificial cognition, and physical action. "Never ask a roboticist what a robot is," Illah Reza Nourbakhsh of Carnegie Mellon (and another expert in autonomous robots) tells us. "The answer changes too quickly. By the time researchers finish their most recent debate on what is and isn't a robot, the frontier moves on as whole new interaction technologies are born."[1] Rodney Brooks, when he was at MIT, called robots "artificial creatures."[2] According to a college textbook chosen at random, "The Robot Institute of America defines a robot as a reprogrammable, multifunctional manipulator designed to move materials, parts, tools, or specialized devices, through variable programmed motions, for the performance of a variety of tasks. *This definition does not exclude human beings.*"[3]

Cynthia Breazeal at MIT gained fame for her work on Kismet, a special variety of humanoid robot that interacted with people using gestures, facial expressions, and sounds. "What is a sociable robot?" she asks in her book on the topic. "It is a difficult concept to define." After pointing to science fiction for some examples, she asserts, "In short, a sociable robot is socially intelligent in a humanlike way, and interacting with it is like interacting with another person. At the pinnacle of achievement, they could befriend us, as we could them."[4] Once again, a leading roboticist resorts to describing robots by what they *do* rather than what they *are*.

Two relatively recent definitions illustrate the lack of consensus, a real problem when it comes to fostering intelligent debate over a topic with far-reaching implications for personal and public life. Maja J. Mataric of the University of Southern California published *The Robotics Primer* in 2007 as a K–12 guide to this important field. Almost immediately in the opening chapter, she states that "a *robot* is an autonomous system which exists in the physical world, can sense its environment, and can act on it to achieve some goals." She goes on to underline her conviction: "True robots ... may be able to take input and advice from humans, but are not completely controlled by them."[5]

Contrast this strict definition, which excludes many familiar machines such as surgical robots, drone aircraft, and industrial robots, with the definition used on a *60 Minutes* broadcast in 2013 that focused on technological unemployment. Narrator Steve Kroft began the segment by saying that "everyone has a different idea of what a robot is and what they look like, but the broad universal definition is a machine that can perform the job of a human. They can be mobile or stationary, hardware or software, and they are marching out of the realm of science fiction and into the mainstream."[6] Apart from the nearly complete lack of overlap with roboticists' definitions, this characterization is notable for its implied theme of machines out of control, rising up against their human masters.

From the core of computer science comes a similarly inclusive reading. Vinton Cerf, widely known for his work on the Transmission Control Protocol (TCP) and Internet Protocol (IP), which underlie all Internet traffic, was elected president of the Association of Computing Machinery in 2012. In an editorial in January 2013, Cerf posited "that the notion of robot could usefully be expanded to include programs that perform functions, ingest input and produce output that has a perceptible effect." After mentioning high-frequency stock trading as an example, he argued that "one might conclude that we should treat as robots any programs that can have real-world, if not physical, effect." In his conclusion, Cerf reveals a broader concern for the implications of contemporary computing and communications, including robotics strictly defined:

"I believe it would be a contribution to our society to encourage deeper thinking about what we in the computing world produce, the tools we use to produce them, the resilience and reliability that these products exhibit *and the risks that they may introduce.*"[7]

This thinking helps move the discussion of robotics away from anthropomorphism and toward instrumentality. For Cerf, a robot is largely a function of software rather than the box it inhabits. But that software increasingly does have implications for the world of atoms, whether in the form of denial-of-service attacks, cyber warfare on physical infrastructure (such as Stuxnet, a computer virus that disabled centrifuges enriching atomic weapons material in Iran), or inside a mobile and potentially autonomous robot.

For our purposes, the most useful definition comes from George Bekey, who wrote in 2005 that "a robot [is] *a machine that senses, thinks, and acts.* Thus, a robot must have sensors, processing ability that emulates some aspects of cognition, and actuators."[8] Robotics is the collected sciences that combine to study, design, and build these devices: computer science is in the forefront, drawing also on materials science, psychology, statistics, mathematics, and various disciplines in physics and engineering. Artificial Intelligence concerns itself with the re-creation of human cognition in silicon semiconductors, either generally or in a delimited domain to be constrained and optimized.

Historical Automata

For all of this definitional uncertainty, billions of people around the world know a robot when they see one, based on literary and Hollywood portrayals, so it is important to note the precise origins of the term. People have been creating automated models of living systems for millennia. Cuckoo clocks, toys, and elaborate automaton hoaxes date back hundreds of years. One of these dates from 1770 and involved a chess-playing automaton with a chess master hiding inside that defeated both Benjamin Franklin and Napoléon I. It provided the name for Amazon's Mechanical Turk, a service in which computers ask people for help in doing tasks computers aren't good at, such as image recognition. A more colorful invention came from France, a country that continues to help define the robotics horizon. A child of the Enlightenment, Jacques de Vaucanson attempted to apply the notions of the clockwork universe to biological life. In 1735, at the age of 26, he invented "some machines that could excite public curiosity," specifically a mechanical duck.[9]

Vaucanson's duck combined a realistic exterior with some unsurprising mechanical characteristics: it could sit, stand, waddle, quack, drink water, and eat pellets of corn. A secondary trick made Vaucanson a celebrity and gained him election to the prestigious Academie des Sciences, alongside René Descartes, Jean-Baptiste Colbert, and Blaise Pascal: the mechanical duck also defecated. People lined up to see the wonder, paying admissions fees equal to a week's wages for some. Vaucanson then became director of France's silk mills, where in 1745 he invented a punch-card system for controlling weaving patterns— a foundation invention for the Jacquard loom, developed in 1801, that proved to influence the early history of computing, when punched cards were once again used to control a sequence of operations. As for the duck, forty years later it was found, like the chess machine, to be a hoax: the bird did no digesting but rather captured the inputs in one reservoir and dispensed the output from a second one.

As the noted policy analyst P. W. Singer points out, Vaucanson's duck provides a vivid illustration of humanity's long-running efforts to create artificial life. In Jewish folklore, the concept of a golem, which is an anthropomorphic figure made from inanimate material, dates back to biblical times. For centuries, the word "android" was used to describe "automatons resembling a human being," to quote the *Oxford English Dictionary*. (The word dates from 1728, less than a decade before the infamous duck.) Mary Shelley published *Frankenstein*, often cited as the first science fiction novel, in 1818 and described the dire consequences of trying to create life in the laboratory. In 1822, Charles Babbage built his "difference engine," a mechanical calculator containing more than 25,000 parts. So people have been trying to create simulations of biological life for a long time. Where did robots enter the picture?

Automata: Science Fiction

By the time he published his play *R.U.R.* (Rossum's Universal Robots) in 1920, Karel Čapek was a well-known Czech intellectual. Like other writers, he was appalled by the carnage wrought by the mechanical and chemical weapons that marked the Great War as a departure from previous combat. The play introduced the robot, an artificial human made from biological material, into the English language as a protest against the dehumanization caused by the modern age: conformity, lack of aspiration, and a cheerful attitude toward dull work were all critiqued with the concept. The word derives from the Czech word "robota," or forced labor, as done by serfs. Its Slavic linguistic root, "rab," means "slave." Thus the original word for robots more accurately defines androids in that they were neither metallic nor mechanical; several were mistaken for human beings in the play.

The play was widely performed and the text translated into many languages. Probably because the robot was an idea whose time had come, for science fiction at any rate, popular depictions proliferated in the following decades.

Thus the definitional dilemma in which we find ourselves dates back about 100 years, to a time when a literary artist used a slavery metaphor to protest the devaluation of human life in a mechanical era. Through the 1920s, robot images centered on the hubris of humans creating life, connecting mechanical invention to *Frankenstein* and earlier legends of overreaching humanity. By 1942, however, the next wave of definition was taking the notion of a robot in a much more positive direction.

Either directly through his writing or indirectly through his wide influence, Isaac Asimov, born Isaak Yudovich Ozimov in Russia in 1920, is single-handedly responsible for most North American conceptions of what a robot can be. Early in his astonishingly prolific career as a writer—with more than 500 books to his credit—Asimov helped found the modern genre of science fiction; in later years, he published literary criticism, nonfiction science writing, mysteries, and novels. After earning the first of his three degrees in chemistry from Columbia in 1939, Asimov teamed up with John W. Campbell, who edited *Astounding Science Fiction* magazine, to generate the Three Laws of Robotics, which both served as the governing principles for his robotic science fiction and guided generations of roboticists whose field at the time lacked formal definition, ethical codes of conduct, and other marks of professional maturity. In the early years of this complex work, science fiction was both powerfully inspirational and the only widely available resource, and Asimov stood at the forefront.

As he later wrote in a nonfiction introduction to the state of robotics in the 1980s, Asimov was "tiring of robots that were either unrealistically wicked or unrealistically noble, [and] began to write science-fiction tales in which robots were viewed merely as machines, built, as all machines are, with an attempt at adequate safeguards."[10] Nine of the stories written in the 1940s in response to this impulse were collected in the book *I, Robot*. In one of them, Asimov coins the term "robotics" and so named an entire discipline of modern science and engineering.

Asimov's Three Laws of Robotics worked well to define a fictional environment but have been less useful in actual practice, no matter how often people have tried to encode them in hardware nearly seventy-five years after they were written as *the premise of a fantasy*. The laws are as follows:

1. A robot may not injure a human being, or, through inaction, allow a human being to come to harm.

2. A robot must obey the orders given it by human beings except where such orders would conflict with the First Law.

3.   A robot must protect its own existence as long as such protection does not conflict with the First or Second Law.[11]

Later, when his stories included robots interacting with entire civilizations and not only individuals, Asimov added a fourth law, called the "Zero Law" in that it came first in logical priority. It took precedence over his previous laws:

0.   A robot may not harm humanity, or, by inaction, allow humanity to come to harm.

Asimov's laws continue to exercise substantial influence in the robotic community, although even a cursory reading suggests they are difficult or impossible to engineer into silicon circuitry. P. W. Singer addresses Asimov's laws in the context of drone aircraft and other military technologies that by definition ignore law number 1. Singer states that the absence of technology-specific ethical codes is troubling in the domain of combat (can a robot be used as an instrument of torture, for instance?) but unsurprising given how most industries regulate drugs, guns, automobiles, and other ethically complex technologies: as lightly as possible.[12] Rodney Brooks, for years the director of the MIT robotics initiative, says quite simply, "we do not know how to build robots that are perceptive enough and smart enough to obey these three laws," adding that it's possible Asimov "did not realize just what a perceptual load these laws put on a robot."[13]

As recently as 2009, professors Robin Murphy of Texas A&M and David D. Woods of Ohio State put forth what they called the "Three Laws of Responsible Robotics." The article was intended as a way to ask necessary questions about responsibility, intention, and unintended consequences, not in fictional stories but real-world factories, nursing homes, and labs. Their laws follow; note the primacy of human responsibility rather than compumechanical wisdom:

1. A human may not deploy a robot without the human-robot work system meeting the highest legal and professional standards of safety and ethics.

2. A robot must respond to humans as appropriate for their roles.

3. A robot must be endowed with sufficient situated autonomy to protect its own existence as long as such protection provides smooth transfer of control which does not conflict with the First and Second Laws.[14]

In short, one problem with the pervasiveness of science fiction is its emphasis on the robot rather than the human as the relevant moral actor.

One problem with
the pervasiveness of
science fiction is its
emphasis on the robot
rather than the human
as the relevant moral
actor.

Gray Area

After more than a half-century of field trials of computer-enabled robots, we still lack a complete or nuanced understanding of what a robot is or is not. In the case of an automated guided vehicle (AGV) following a strip on the floor of a warehouse, there is sensing and locomotion, but it would appear that cognition can be absent, or minimal. Notably, the first AGVs were not called "robots," and even today, it is unclear how such devices are categorized.

By contrast, the industrial robot—typically anchored in place, doing a repetitive task by electronic memory, and operating inside a cage for human safety—can trace some of its U.S. origins directly to Asimov. Joseph Engelberger was explicitly motivated to build something much more than a new kind of machine tool:

> Over and over, the advice was, "Don't call it a
> robot. Call it a programmable manipulator. Call it a
> production terminal or a universal transfer device."
> The word is *robot* and it *should* be *robot*. I was
> building a robot, damn it, and I wasn't going to
> have any fun, in Asimov's terms, unless it was a
> *robot*. So I stuck to my guns.[15]

The sense-think-act paradigm proves to be problematic for industrial robots: some observers contend that a robot needs to be able to move; otherwise, the Watson computer might qualify. Another more recent example comes from Silicon Valley. The Nest is a learning thermostat. Powered by sensors and connected by Wi-Fi, it tracks a household's behaviors and modifies the temperature automatically. The team behind the Nest has more than its fair share of advanced degrees in computer science and robotics, and a clear record of genuine consumer product innovation: many worked on the Apple iPod or Google search engine. The Nest senses: movements, temperature, humidity, and light. It reasons: if there's no activity, nobody is home to need air conditioning. It acts: given the right sensor input, it autonomously shuts the furnace down.

Fulfilling as it does the three conditions, is the Nest, therefore, a robot? (The fact that Google bought the start-up at about the same time it made other robot investments suggests to some people that it is.)

Intuitive Surgical makes the da Vinci Surgical System. Using sensitive joysticks, a surgeon makes the da Vinci move probes and surgical tools inside a patient's body. The system certainly has sensors, and it acts on human patients. But without autonomous cognition, can the da Vinci really be called a "robot"?

Asimov did not define what a robot was but posited a hypothetical moral system to which idealized robots should adhere. Dictionaries are of no help. Hollywood portrayals will be analyzed further, but, for now, suffice it to say that neither the 1960s cartoon character Rosie Jetson nor Stanley Kubrick's HAL nor George Lucas's R2D2 defines what a robot is or is not. Neither does any one of the hundreds of industrial robots at work in a typical auto factory. Yet almost all of us know a robot when we see one.

ROBOTS IN POPULAR CULTURE

Mythos

An ongoing history of efforts to create artificial life—to replicate, enhance, or surpass human characteristics—dates back thousands of years to the golem, with cuckoo clocks, Mechanical Turk, and Vaucanson's duck following later. What could motivate such a persistent quest? Humans might have sought validation as creators, to be on a level with whatever deity or other maker their religion names. Robert Geraci, a religious studies professor, suggests another explanation: the story of Adam and Eve is deeply ingrained in Western culture. The myth goes on to posit that humanity, from that time forward, has lived in a state of having fallen from grace. Thus the quest for artificial life can be seen as an attempt to escape that imperfection, and potentially bring about a new era.[1]

And the quest is not merely Judeo-Christian. Respected Japanese robot designer Hiroaki Kitano stated the case in strikingly similar terms—his naming of the robot's origin is critically important here—suggesting that the mental model is widely held: "In his gestation, [the humanoid robot] PINO symbolically expresses not only our desires but humankind's frail, uncertain steps toward growth and the true meaning of the word human."[2]

Technology has long been viewed as both an impediment (by Shakers and Amish among others) and an aid to a higher state of being. Geraci identifies the "digital utopians" of the homebrew computer age who thought the decentralization of computing and thus knowledge could help bring about flatter, more egalitarian power structures. He then traces the idea of "apocalyptic AI" through a variety of expressions, including transhumanists who see artificial life as a step beyond humanity's imperfect (they would not call it "fallen") state. Science fiction and popular scientific writing blend in amplifying this theme. The roboticist/novelist Hans Moravec writes of "mind children" (AI robots) defeating humans in a Darwinian struggle,[3] while Ray Kurzweil uses characters from the future in several books to explain his notion of the "Singularity."

Deep religious concepts, including salvation, eternal life, and some state of otherworldly perfection are all informing our discussions of robots just as surely as are considerations of battery life, machine vision, and path-planning algorithms. As Geraci asserts, "The sacred categories of Jewish and Christian apocalyptic traditions have thoroughly penetrated the futuristic musings of important researchers in robotics and artificial intelligence." Given that these ideas have emerged in disparate areas—online gaming, popular culture, all sorts of computerized user interfaces, autonomous vacuum cleaners, drone warfighting, high-frequncy stock trading, automated factories—the study of robots provides a fascinating keyhole into contemporary existence. "To study intelligent robots," Geraci concludes, "is to study our culture."[4] Going one step farther, the Carnegie Mellon roboticist Illah Reza Nourbakhsh argues that more than culture is involved: "the robotics revolution can affirm the most non-robotic quality of our world: our humanity."[5]

The other mythology we should acknowledge derives from the place of technology in American culture, dating from the earliest days. The history of the United States is unique in several regards: its relationship to Europe, its sheer size, its vast mineral wealth, and its contestation. In no other country has there been a replacement of indigenous people at such scale, by such a wide variety of immigrants. A core piece of Americana is the role of a physical frontier, the westward-moving demarcation between "civilization" and the unknown. Even after California and the interior West were populated, the idea of a dividing line waiting to be explored and colonized remained potent. The Apollo moon landing was situated squarely in this narrative, and then science fiction even more aggressively incorporated the idea, most straightforwardly in the opening to nearly every *Star Trek* television show ever broadcast:

"Space: the final frontier. These are the voyages of the starship *Enterprise*. Its five-year mission: to explore strange new worlds, to seek out new life and new civilizations, to boldly go where no man has gone before."[6]

The conquest of the North American frontier, and the inhabitants thereof, was largely accomplished with technological innovation: rifles, railroads, barbed wire, and the telegraph in the nineteenth century, followed by irrigation, air-conditioning, and interstate highways in the twentieth.[7] Technology helped establish a physical frontier, then itself became a metaphorical frontier (serving many of the same purposes) once the territory was settled.

In mythological terms, frontiers can apply neatly to science, knowledge, and innovation: a Google search of "frontiers of science" returns 31 million hits. Although many countries have active robotics research programs, among them Japan, France, Germany, and Sweden, the U.S. robotics effort should be situated in proximity to this rich mythological past of unique ideals: conquest, expansionism, and one other trait. For lack of a better word, let us call it "solutionism."

In one reading, this neologism refers to a particularly U.S.-centric attitude, often accompanied by some naïveté, holding that most problems have solutions, often technological ones. More acerbically, the cultural critic Evgeny Morozov calls solutionism "an intellectual pathology that recognizes problems as problems based on just one criterion: whether they are 'solvable' with a nice and clean technological solution at our disposal."[8] Whichever reading you accept, there is something in the U.S. ethos that seems to encourage tinkering, rather than a more "realistic" view of the world that accepts that some things just cannot be fixed.

With these two long-term backdrops in place, let us turn to the place of robots in Western culture, primarily science fiction literature, cinema, and television. It's difficult to recall an emerging technology with deeper roots in science fiction than robotics. From the very origin of the word, the history of robotics coincides with, and was heavily shaped by, images and legacies of books and movies. To the extent that science fiction is a relatively young genre, and that cinema and television are young media, this path of cultural influence is unprecedented. Its consequences, however, are substantial and largely invisible. As robotics becomes a much more feasible and familiar field, understanding its cultural origins becomes a necessary step in defining what robots are, what humans want, and how the two entities interact.

R.U.R.

R.U.R., the play that introduced the word "robot," was a critique of mechanization and the ways it can dehumanize people. It premiered in Prague in 1921 and was one of the most performed plays of the twentieth century, being translated and performed in "almost all the civilized countries of the world," as one 1962 analysis put it.[9] As its author Karel Čapek told a magazine,

> The old inventor, Mr. Rossum (whose name translated into English signifies "Mr. Intellectual" or "Mr. Brain"), is a typical representative of the scientific materialism of the last [nineteenth] century. His desire to create an artificial man—in the chemical and biological, not mechanical sense—is inspired by a foolish and obstinate wish to prove God to be unnecessary and absurd. Young Rossum is the modern scientist, untroubled by metaphysical ideas; scientific experiment is to him the road to industrial production. He is not concerned to prove, but to manufacture.[10]

Thus the very word now so much in general circulation was a cultural critique of both the human desire to replicate his own form, much in the manner of Dr. Frankenstein, as well as the logic of industrial production.

The contrast between robots as mechanical slaves and potentially rebellious destroyers of their human makers echoes *Frankenstein* and helps set the tone for later Western characterizations of robots as slaves straining against their lot, ready to burst out of control. The duality echoes throughout the twentieth century: Terminator, HAL 9000, *Blade Runner*'s replicants. The character Helena in *R.U.R.* is sympathetic, wanting the robots to have freedom. Radius is the robot that understands his station and chafes at the idiocy of his makers, having acted out his frustrations by smashing statues.

Helena: Poor Radius. … Couldn't you control yourself? Now they'll send you to the stamping-mill. Won't you speak? Why did it happen to you? You see, Radius, you are better than the rest. Dr. Gall took such trouble to make you different. Won't you speak?

Radius: Send me to the stamping-mill.

Helena: I am sorry they are going to kill you. Why weren't you more careful?

Radius: I won't work for you. Put me into the stamping-mill.

Helena: Why do you hate us?

Radius: You are not like the Robots. You are not as skillful as the Robots. The Robots can do everything. You only give orders. You talk more than is necessary.

Helena: That's foolish Radius. Tell me, has any one upset you? I should so much like you to understand me.

Radius: You do nothing but talk.

Helena: Dr. Gall gave you a larger brain than the rest, larger than ours, the largest in the world. You are not like the other Robots, Radius. You understand me perfectly.

Radius: I don't want any master. I know everything for myself.

Helena: That's why I had you put into the library, so that you could read everything, understand everything,[11] and then—Oh, Radius, I wanted to show the whole world that the Robots were our equals. That's what I wanted of you.

Radius: I don't want any master. I want to be master over others.[12]

Helena's compassion saves Radius from the stamping mill, and he later leads the robot revolution that displaces the humans from power. Čapek is none too subtle in portraying the triumph of artificial humans over their creators:

Radius: The power of man has fallen. By gaining possession of the factory we have become masters of everything. The period of mankind has passed away. A new world has arisen. ... Mankind is no more. Mankind gave us too little life. We wanted more life.[13]

Humans were doomed in the play even before Radius led the revolt: when mechanization overtakes basic human traits, people lose the ability to reproduce. As robots increase in capability, vitality, and self-awareness, humans become more like their machines: humans and robots, in Čapek's critique, are essentially one and the same. The measure of worth, industrial productivity, is won by the robots that can do the work of "two and a half men." Such a contest implicitly critiques the efficiency movement that emerged just before World War I, with its time-and-motion studies, which ignored many essential human traits.

The debt of *R.U.R.* to Shelley's *Frankenstein* (1818) is substantial, even though the works are separated by almost exactly a century. In both cases, humans show hubris by trying to create artificial life. (Recall that even today, Rodney Brooks refers to robots as "our creatures.") Whether humans get the recipe wrong, as in the earlier novel, or make beings smarter than the humans who spawned them, as in the case of Čapek's play and its offspring, humans pay the price for aspiring to play God. In both works, the flawed relationship between creator and creature drives the plot, and in both cases, the conflict ends in bloodshed.

Few people today know *R.U.R.* More know Asimov's robot novels, but in either case, robotics is a special branch of technology in that, in the archetype, it aspires to approximate humanity, whether through biomimicry, flawless logic, or economic superiority.

Robots at the Movies (and Elsewhere)

This dualism between robots as slaves and robots as potential overlords, rebelling against the constraints of servitude, plays out remarkably consistently in the twentieth-century West. Only six years after the premiere of Čapek's play, the German director Fritz Lang released *Metropolis*. The work remains widely recognized as one of the most influential movies ever made, not least because it contributed heavily to Nazi Party ideology. Many themes carry over from *R.U.R.* including confusion among humans as to the true identity of a robot among them, a looming revolt of the dehumanized industrial workers, and romantic attraction between human and robotic species. In the end, the mechanical Maria (who has been instigating the workers to rise up and destroy their machines) is discovered and burned at the stake, while the robot's human counterpart escapes after being kidnapped and helps broker a peaceful resolution between workers and factory owners. The silent film originally ran about two and a half hours, took 310 days to shoot and involved 36,000 extras. Despite its length and confusing plot, it remains a film landmark. The portrayal of the robotic Maria as deceitful and rebellious tracks closely to the prototype established by Čapek.

The next memorable character appeared in 1939 in one of the most beloved films in cinematic history. The portrayal of the Tin Man in *The Wizard of Oz* aligns closely with other robot figures we will encounter in that he is the result of a series of mechanical prosthetics installed when, as woodsman, he injures each of his limbs, one by one; but the tinsmith, in giving him also a tin body, forgets to give him a heart. The Tin Man escapes the Frankenstein associations in part because of the incremental nature of his mechanization, and because the focus of the film (and the book from which it was made) is not on his strength or logic, but on his desire for feelings. The character has become an icon, appearing in other novels, pop songs, and advertising campaigns decades after the movie's release.

Also in 1939, at the New York World's Fair, Westinghouse demonstrated a 7-foot aluminum-skinned robot named "Elektro." The robot could smoke cigarettes, count on its fingers, and speak by way of a 78 rpm phonograph record embedded inside. A year later, the robot was joined by a metal dog named "Sparko" that could sit, beg, and bark. Elektro was recently restored and resides in the Mansfield, Ohio, Memorial Museum, having been built at Westinghouse's Mansfield facility.

Roughly a decade after *Metropolis*, science fiction as a genre was in the midst of both a growth in popularity and the development of robots as critical elements of both plot and theme. Although the arrival in the 1940s of what became known as the "big three" of science fiction—Arthur C. Clarke, Robert Heinlein, and Isaac Asimov—helped establish the genre as a marketable and recognizable body of work, it was in the writings of Asimov that robots gained new status and robotics was foreseen as a discipline alongside mechanics or dynamics. In a series of short stories written between 1939 and their collected publication in *I, Robot* in 1950, Asimov introduced the notion of a "positronic brain," a computer that operated at a sufficiently high level that artificial beings could express a consciousness recognizable to humans.

Rather than inhabit the *Frankenstein* stereotype of artificial beings scheming to destroy their creators (what Asimov called "Robot-as-Menace"),[14] robots in Asimov stories consistently exhibit moral codes that were explored by way of conflicts rather than mere exposition. In the story "Runaround," published in 1942, Asimov stated the three laws of robotics, squarely in line with his emerging sense that robot stories needn't be either threatening or sentimental. Instead, he wrote in 1982, "I began to think of robots as industrial products built by matter-of-fact engineers. They were built with certain safety features so they weren't menaces and they were fashioned for certain jobs so that no Pathos was necessarily involved."[15] A concept of "robopsychology" in Asimov's work helps human observers understand why robots made the decisions they do, for example. Complex themes such as the value of work, the attractions between humans and robots, and the relative value of a robot life versus a human one also appear regularly.

One primary genre of robotic fictional representation features aliens from space that inhabit mechanical, or biomechanical, forms. The distinctions among robot genres become important here. Robots, culturally rather than technologically speaking, tend to be mechanical entities that exhibit anthropomorphic tendencies. The word "android" dates to the nineteenth century and refers to non–human beings that have flesh-like exteriors. (Strictly speaking, Čapek's "robots" in *R.U.R.* are androids and not robots.) Cyborgs, meanwhile, are a more recent conceptualization, dating to about 1960. The MIT professor Norbert Wiener coined the term "cybernetics" to denote "the scientific study of control and communication in the animal and the machine."[16] Cyborgs are thus beings that merge artificial and organic systems of control. Although the term "cybernetic organisms" can be applied broadly to large systems, cyborgs are for our purposes typically semihuman characters that have been enhanced in some way through the application of computational/robotic capabilities.

When robots descend from outer space in popular culture, they can be any of these species. In *The Day the Earth Stood Still* (1951), an alien named "Klaatu" brings a message promoting world peace (implicitly supporting the creation of the United Nations), accompanied by the robot Gort; the robot emerges from a spaceship designed in collaboration with Frank Lloyd Wright. In 1956, Robby the Robot appeared in *Forbidden Planet* alongside future *Airplane* star Leslie Nielsen. After the original movie, Robby later appeared in other films and TV shows. Significantly, humans traveled to outer space where they encountered Robby, rather than the other way around. After 1960, humans' adventures in space became a more common theme, with several landmark films following the plot device.

Space travel proved to be one of several convenient retellings of old tales in the robot genre. The debt of many movies to Homer (including of course *2001: A Space Odyssey*) is obvious. The Swiss Family Robinson of 1812 (while sailing from Europe to Australia, they were shipwrecked in the East Indies) paid its debt to *Robinson Crusoe*, then itself became the inspiration for a 1960s television show. Fathers and sons (*Star Wars*), coming of age, mistaken identity, and of course the Frankenstein monster threat all make repeated appearances, with the twist that robotic characters are only partially human.

In 1963, the BBC launched the *Doctor Who* television serial, in which Daleks featured prominently. These aliens were in fact extraterrestrial cyborgs that mutated into ruthless killers, given that they lacked all emotions save for hatred; their most frequent utterance is simply "Exterminate!" Like other fictional cyborgs, Daleks quickly became a touchstone in British popular culture, and have appeared in a fiftieth anniversary commemoration done by the BBC. *Doctor Who*, meanwhile, remains popular among many populations, including some U.S. teens.

The depictions of robots in popular culture began to shift after Asimov. In contrast to the nonhuman assemblies of mechanical parts typified by Gort or Robby, later robots could express a variety of emotions, interact with human characters with nuance and ambiguity, and wrestle with the complexities of their nature. Although there continued to be clear-cut evil villains, and simplistic servants, the commercial success of multiple robot-themed movies indicated a growing recognition within mass culture of the (largely fictional) possibilities of a new technology and potentially new life-form.

Although alien movies continued to appear after 1970, more and more cyborgs were situated on Earth. In *Robocop* (1987), for example, a murdered Detroit police officer is returned to life with electromechanical spare parts and given superhuman powers. Not surprisingly given the genre, the robotic creature's powers create tensions with human laws and networks of corruption; the threat of a rogue, nearly invulnerable humanoid echoes other narratives. *Robocop* was far from original, following as it did director James Cameron's robot-movie milestone, *The Terminator* (1984). Made for roughly $6.5 million, the film grossed $78 million (more than $225 million in 2015 dollars). Arnold Schwarzenegger played a ruthless cyborg, with flesh-like skin over a bionic endoskeleton. The fate of humanity hung in the balance as an unborn baby (John Connor, whose initials are obvious), who was to lead a humanist uprising, was being stalked by Schwarzenegger's character, time-traveling from the future in the employ of the evil Skynet.

In *Blade Runner* (1982), director Ridley Scott used yet another space-travel device to create a dystopic vision of the future based on the Phillip K. Dick novel *Do Androids Dream of Electric Sheep?*[17] Cyborgs are created on Earth by evil corporations, and then shipped off to work on space colonies. Some sneak back to Earth, in violation of the law, so they must be hunted down. Known as "replicants," the androids represent yet another vision of robots as capable of dull, dangerous, and dirty work, but also capable of desiring to transcend their station and challenge humans for authority. The character played by Daryl Hannah continues the trope embodied by Maria in *Metropolis*, the alluring female android who preys on men's weakness.

What countless polls and surveys have called history's greatest space movie, however, features a robot that never appears. Arthur C. Clarke, one of science fiction's "big three," collaborated with Stanley Kubrick on *2001: A Space Odyssey* (1968), which portrayed robots in yet another light. Rather than assuming anything resembling human form, HAL 9000, the computer that runs the spaceship bound for Jupiter, has a human-sounding voice, appears to express emotions, can sense its environment (including lip-reading the astronauts when they express concern about HAL's behavior), and can also operate the ship's systems and devices. HAL cannot stop the remaining human astronaut from disconnecting its power supply in the film's crucial scene after HAL has killed all the other astronauts, and exhibits emotional regression as its computational function diminishes. In Clarke and Kubrick's portrayal, a robot is thus capable of emotion, highly intelligent, and powerful (though not all-powerful), but ultimately proves it cannot be trusted.

The film was a massive commercial success, and remains a powerful cultural monument nearly fifty years later. Ambiguity—probably the film's defining characteristic—allows viewers to re-create in HAL their personal fears, aspirations, and associations of artificial intelligence. The fact that HAL is never seen, for example, but only heard through the excellent voicing of Douglas Rain, contributes much to the film's timeless quality. Every serious robot movie that came after, and many not-so-serious ones, self-consciously relates itself to this monument.

At the opposite pole of *2001*'s HAL are multitudes of robot characters that embody neither menace nor superior intellect. From the cartoon maid Rosie on *The Jetsons* (aired in primetime in 1962–63 and in other time slots until 1987), to Wall-E, the protagonist of the 2008 Disney/Pixar film by that name, robots often proved to be useful devices for producers and directors: they could inject comic incongruity or address more serious topics at arm's length. The robots' voices and gestures could satirize human traits, but humanity was portrayed as a superior state given how many robots aspired to it. At the same time, robots are frequently portrayed doing menial tasks as at least part of their identity, implicitly freeing humans from drudgery.

B9, the faithful "environmental control" robot on the U.S. television series *Lost in Space* (1965–68), became famous for warning the "Space Family" Robinson's son Will of impending harm: even today, "Danger, Will Robinson!" is still uttered by people unaware of the original context. But by far the apotheosis of the robot-as-faithful-servant model is the Laurel-and-Hardy duo of R2D2 and C-3PO from the *Star Wars* franchise. Indeed, the shadow cast by the George Lucas creations is so long that many other films must characterize their robots in contrast to the much-loved pair of metallic icons.

R2D2, the shorter of the two, was given part of its appeal by a human actor, Kenny Baker, who controlled its movements from inside during filming. (Other shots featured a radio-controlled, uninhabited version of the robot.) The robot speaks in a machine language, translated by his compatriot, adding to the comedic dynamic, especially when the jokes are on C-3PO. With both characters, there is no possibility of robot rebellion; potentially because of the convincing nature of the futuristic space-age setting, there is no hint of human hubris at work, possibly because the robots could have otherworldly origins.

As for C-3PO, the brilliant butler-esque voicing by the English actor Anthony Daniels contributes enormously to the character's appeal. The superhuman characteristics common among cinematic robots are in this instance cerebral rather than mechanical: even though it can translate "six million forms of communication," C-3PO is in fact a coward in the rebels' battles. Its physical form is highly derivative of Maria's metallic shell in *Metropolis*, while the dynamic between the robots is as familiar as any odd-couple or buddy movie.

Cultural Signals

Taken as a whole, what do we make of robots in Western popular culture?

1. The directors who have advanced the cultural representation of robots include some of the industry's leading figures: Woody Allen (*Sleeper*), James Cameron (*Terminator*), Chris Columbus *(Bicentennial Man),* Stanley Kubrick, George Lucas, Ridley Scott, and Steven Spielberg (*A.I. Artificial Intelligence*). These men have made movies collectively grossing tens of billions of dollars. For them to share the use of robots as key characters suggests the massive cultural appeal of the technology/archetype.

2. Any generalizations about movie and literary characters have little in common with real robots, whether hardware (Boston Dynamics' Atlas) or software (Wall Street high-frequency trading systems): robots cannot feel pain, or understand when they inflict it. Robots do not have cognition that differentiates between humans and other mammals, or between humans and other things that move. Robots have no mechanism for aspiration. Robots, however autonomous, depend on humans for everything from battery power to software updates. Robots cannot choose except among predefined choices; they cannot "consciously" subvert their creators because robots lack self-awareness at the cognitive level.

3.   The high bar for real robots to do such things as open doorknobs, traverse uneven terrain, or perform logic at the level of a child features only minimally in science fiction and cinema. Thus when people see hard tasks accomplished within the field of robotics (such as a robot folding a shirt or opening a bottle of beer), it appears underwhelming. Many observers put substantial emphasis on Asimov's three laws when they bear little resemblance to real robotic science. The cognitive scientist Steven Pinker puts it in stark terms: "When Hamlet says, 'What a piece of work is man! how noble in reason! how infinite in faculty! in form and moving how express and admirable!' we should direct our awe not at Shakespeare or Mozart or Einstein or Kareem Abdul-Jabbar but at a four-year-old carrying out a request to put a toy on a shelf."[18] AI has a long way to go to approximate such responses.

4.   The distinctions among robots, androids, and cyborgs are largely immaterial to audiences. Aliens, for example, are equally plausible elements in these plots.

5.   Thus the images of robots widely present in mass culture do little to help communities confront real issues of emerging technologies. Because perceptions are highly conditioned by the robot-as-menace, robot-as-pathos, robot-as-self-aware, or robot-as-servant/butler stereotypes, substantive discussions of prostheses, to take one example, have been slow to form.

To summarize, Čapek's conception of a robot, a Czech word, embodied a metaphor of humanity as slaves to technology, with the danger that the artificial creation might rebel against its human creators much as Frankenstein's creature did. About twenty years later, Isaac Asimov created literary robots that embodied human traits, to the point where "robopsychology" provided insight into their workings. Finally, the notion of artificial intelligence, however embodied or disembodied, is often conflated with superhuman logic that will eventually outstrip humans' powers of thought and feeling. For all their differences, HAL 9000 and IBM's Watson function similarly as culturally frightening figures: their superhuman capabilities can apparently nullify human capabilities and designs. In each of these cases, the power of cultural iconography can distract us from concrete, live issues: who besides Stephen Hawking should have robotic augmentation? Should labor unions invest in robots that replace human workers? Who can implement and override safety interlocks for a given class of robot? And so on.

An Alternative: The Japanese Tradition

Note that all of the aforementioned cultural resources were explicitly Western. From *Frankenstein* through *Star Wars* and *The Terminator*, the theme of "technology out of

The notion of artificial intelligence, however embodied or disembodied, is often conflated with superhuman logic that will eventually outstrip humans' powers of thought and feeling.

control," as the academic Langdon Winner put it, is always present: created life often turns out to be, as Asimov said, "robot-as-menace." But it is impossible to assess the current state of robotic science and representation without including the Japanese.

Although *manga*, the distinctive Japanese comic art form, dates from the late nineteenth century, it can trace its origins to thirteenth-century scrolls. After World War II, when the nation was reforming its mythology in light of both military defeat and foreign cultural influences brought with the U.S. occupation, manga emerged as a vehicle for heroism, virtue, and prowess. One artist, Osamu Tezuka, created a character that captured his nation's imagination. Called "Mighty Atom" in Japan, he is more familiar to Western audiences as Astro Boy; one advisor thought the atomic connotations might be problematic for U.S. audiences after Hiroshima and Nagasaki. Mighty Atom has become an essential part of Japanese culture: he actually has full Japanese citizenship; Tezuka is regarded as "the god of comics," on the same plane as Walt Disney, but with elements of Arthur C. Clarke, Stan Lee, Tim Burton, and Carl Sagan also in the mix, according to his biographer.[19]

Tezuka was a fascinating character. Born in 1928, he developed his considerable artistic talent in part by drawing skillful catalogs of insects as a teenager. He later graduated from medical school (where the drawings in his lab notebooks were exquisite) but did not practice medicine. His various characters, led by Mighty Atom, were organized into a "star system" modeled on a Hollywood movie studio. They generated a massive following. Tezuka then developed his own production team that eventually made the first TV animations in Japan. Different characters and series, including his most ambitious story called "Phoenix," helped turn manga into a multibillion dollar industry, both in Japan and worldwide. Tezuka's work was so skillful, innovative, and compelling that Stanley Kubrick asked him to be the art director of *2001*, but Tezuka could not commit to move his team to England for a year during production. Tezuka died of stomach cancer in 1989, still drawing until his last day.

Mighty Atom originally appeared as a supporting element in 1951 but was featured as a title character in 1952. He appeared regularly until the 1970s, and occasionally in the years after he had become a national icon. His fictional date of birth, April 7, 2003, was marked by nationwide celebrations; four years earlier, the character had been used to help launch Toyota's Prius hybrid vehicle in Japan. Through multiple reinforcing channels, the robot's likeness is a familiar image in Japanese culture.

Mighty Atom's life originates in tragedy. When Dr. Tenma, the head of the Ministry of Science lost his son Tobio in a collision between a flying car and a heavy truck, he convenes experts and builds a robot in Tobio's likeness.

When he discovers that the robot cannot mature and cannot learn to love natural beauty, Tenma gives the robot to a circus. Tenma's successor at the Ministry of Science, Professor Ochanomizu, finds the robot, discovers he can feel emotions, and nurtures him as a foster son. The scientist urges him to use his powers for good, primarily fighting crime but also, in one episode, intervening on behalf of a Vietnamese village before U.S. planes were scheduled to bomb it. Thus the entire story is rooted in the robot's need for human company and acceptance: robot rights are a common issue engaged in the manga.

Mighty Atom has super powers, but not in the *Terminator* vein: he stands 4 feet 6 inches (1.37 meters) and weighs 67 pounds (about 30 kilograms), but packs a 100,000 horsepower (roughly 75,000 kilowatt) atomic engine along with retractable jets in his hands and feet. Over the course of many stories, it is revealed that Mighty Atom has

- Jet-powered flight
- Multilingualism (sixty languages)
- Analytical skills
- Searchlight eyes
- Supersensitive hearing

- Hidden weapons in his back
- The ability to tell if a person is good or evil.[20]

This mix of characteristics, combined with Mighty Atom's diminutive stature, creates a myriad of plot possibilities. He can sweat while solving challenging situations, but must be altered by his foster-father to be able to shed tears. His powers were not infinite: Atom could run out of fuel, and when he ate human foods, they ended up in his machine-filled chest cavity and had to be removed. His enemies included bad people, people who hated robots, rogue robots, and alien invaders. Time travel was a frequent feature of the stories.

Much like Asimov roughly a decade earlier, Tezuka developed a code of robot law to guide Mighty Atom through his adventures:

1. Robots are created to serve mankind.

2. Robots shall never injure or kill humans.

3. Robots shall call the human that creates them "father."

4. Robots can make anything, except money.

5. Robots shall never go abroad without permission.

6. Male and female robots shall never change roles.

7. Robots shall never change their appearance or assume another identity without permission.

8. Robots created as adults shall never act as children.

9. Robots shall never assemble other robots that have been scrapped by humans.

10. Robots shall never damage human homes or tools.[21]

Several aspects of the list deserve mention. Note the similarity of rules 1 and 2 to Asimov's laws; note also that there is no parallel to Asimov's Third Law, that robots should protect themselves. Deception by robots is explicitly forbidden, three times over.

The tensions between Atom's eternal childhood, superhuman powers, moral compass, and desire for human love drive his identity. It both shaped and reflected Japanese attitudes toward robots at large. The Sony Aibo, for example, clearly reflects a sensibility in line with Atom: the appealing, nearly cuddly presentation of robots in many Japanese scenarios (Paro the seal being only one) conveys neither cold utilitarianism nor potentially demonic upheaval.

The contrast can be readily observed when comparing Mighty Atom to a contemporary Japanese manga robot, the giant Iron Man, which began appearing in 1956. Weighing 25 tons and standing 20 meters (66 feet) tall,

Iron Man had been built during the war as a secret weapon but was then turned to peacetime uses. Unlike Atom, he was not autonomous but was remote controlled, usually by a clever young boy, but subject to being turned to evil purposes if the remote were stolen. Although the boy did fight evil, much like Atom, the moral neutrality of Iron Man helped delimit a larger dialectic within Japanese culture: much like that expressed in Asimov's laws, there exists a desire to have robots help humanity rather than hurt it, embodying a higher moral vision than humans can claim for themselves. Given that Iron Man at times eclipsed Mighty Atom in popularity, his presence serves as an extremely potent and useful counterweight to what Asimov might have called the "robot-as-saint" school of thought.

Robots and Myth

Robots are technological tools. Yet few tools have such rich mythologies supporting them. Even more significantly, that mythos largely preceded the actual achievement of most of the milestones of autonomous robotics. The history is both long, dating to biblical times, and hyperreal, with the most recent advances in storytelling being used to advance and shape the narrative. There is a strand of robot-ness that is particularly North American, and another aspect that stands independent of any given culture.

Robots are paradoxical in other ways as well. They serve as slaves, or servants, yet they are feared as potential overlords. Robots absorb many projections of perfection that people cannot attain, yet have trouble executing some basic human maneuvers. Robots offer the prospect of radical advances in leisure while raising questions of what humans will do for a livelihood.

By now, it should be well demonstrated that robots have been introduced into the cultural dialogues of both East and West in ways that differ considerably from past technologies: radios, air conditioners, automobiles, or even smartphones were never portrayed as human creations that desire to eclipse their makers, for example. The place of anthropomorphism in our discussions of what robots are, what they can do, and how they should be understood marks a key point of departure within the technological history of Western countries. All of that notwithstanding, what is the actual state of robotics?

ROBOTICS IN THE PRESENT TENSE

Robotic devices are quietly permeating modern life. In nearly every domain, a wide variety of these technologies can increase accuracy, free humans from danger or drudgery, overcome the limits of human fatigue and limited sensory capacity, and extend human presence. Robots can also dehumanize work and relationships, increase economic disruption, and give rise to other negative consequences for humans. The breadth of activity, variety of contexts, and speed of progress all contribute to a major change in the way people use computing.

Artificial Intelligence

If we use the "sense-think-act" model to define a robot, the "think" component deserves special attention. At the most basic level, artificial intelligence describes efforts to re-create some degree of human reasoning with non-human elements or devices. Although the aspirations date from antiquity, the current landscape navigates from a benchmark placed in 1956, when a conference of early computer scientists at Dartmouth College formalized attempts to create electronic function that mimicked the human brain. John McCarthy (later the director of SAIL, Stanford's AI lab) is credited with coining the term in that year, and Marvin Minsky founded MIT's AI lab in 1957.

Pursuing robotics was difficult in the 1960s and 1970s. Processing machinery was slow and big (the personal computer had not yet been invented), wireless networks were slow and proprietary, and vision systems were slow, expensive, and low resolution. Much effort was devoted to building comprehensive cognitive maps of the robot's environment, to "knowing" the world before interacting with it, but especially given slow processing, this approach produced limited results.

A parallel effort was under way in AI computing. Cyc was founded in 1984 as an attempt to build a comprehensive computer ontology of everything. Having been taught that "rain is a form of water" and that "water feels wet on the skin," the computer could potentially infer that, if I state, "I came in from the rain," then I must be wet. At the time of its founding, the lead scientist claimed it would take 350 man-years to build the rules engine, but thirty years on, it remains unfinished.

The field has weathered periods of extremely generous funding followed by falls from favor. In the 1990s, jokes about the flush "AI '80s" were common in some circles. As funding decreased, refugees from AI dispersed; some went into search while others migrated into genomics and other biomedical fields. After the emergence of Google, and after the Deep Blue chess computer defeated Garry Kasparov in 1997, AI fields returned to prominence and funding, both governmental and investor. One current topic attracting high levels of interest is natural language processing (NLP), popularized in the Apple Siri, Google's type-ahead and other search tools, and IBM's Watson computer. More than merely recognizing voice, natural language processing must disambiguate homonyms ("bass" the fish vs. "bass" the low-frequency sound), understand context ("What's that building off to the right?"), and decode jokes, malapropisms, and other "illogical" statements.

The importance of AI for robotics is obvious: human-computer interaction, physical locomotion, collision avoidance, and image recognition all rely on tools that, at some level, must mimic or replace human cognition.

Industrial Robots

At the same time that computer scientists aspired to make machines more like brains, a separate group of entrepreneurs was looking to replicate human muscle and bone. The story of their combined efforts unfolded far from university labs, in garages and machine shops. In the United States, the two main figures were George Devol and his associate Joseph Engelberger. Devol applied for the first robotics patents in 1954 (predating the Dartmouth AI conference); these were granted in 1961. Devol and Engelberger founded Unimation in the mid-1950s and produced the first industrial robot, the Unimate. It transferred work in process between factory locations several feet apart. Japan entered the market when Kawasaki Heavy Industries licensed the technology from Unimation.

Adoption of robotics technologies was slow in the 1960s: foreign automotive competition was not yet intense, and large manufacturers feared moving out of step with their industry counterparts. As of 1964, Unimation had sold only thirty robots and cash flow was an issue, but between 1967 and 1972, Unimation's cumulative sales volume soared from $2 million to $14 million.

In the mid-1960s, a graduate student named Victor Scheinman designed robotic arms for both Stanford's and MIT's AI labs before taking a fellowship at Unimation to commercialize his ideas. In conjunction with General

Motors, which specified that the new robot move in the same amount of space occupied by a human, with comparable reach, Unimation then brought the PUMA (Programmable Universal Machine for Assembly) to market in the mid-1970s, and the worldwide market for industrial robots took off. ASEA Brown Boveri Corporation (ABB) from Sweden, General Electric, and KUKA from Germany all made serious commitments. General Motors formed a joint venture with Japan's FANUC; Westinghouse bought Unimation for $107 million in 1984 then sold it to a French firm, Staubli, four years later.

Industrial robots are essentially programmable machine tools that perform a sequence of actions, typically in assembly-line scenarios. With just over a million industrial robots installed worldwide, according to the International Federation of Robotics, industry revenues amounted to about $9.5 billion in 2014.[1] After robots were deployed in automobile factories, the number of electronics factories using robots has grown rapidly in the past decade; Foxconn, the Taiwan-based company that assembles Apple products and similar goods, announced its intention to install a million robots—in one company alone—after 2012.[2] Despite China's comparatively low wages, future projections favor the economics of a three-shift robot that does not sleep late or come to work having a bad day; that does not require breaks, heating or cooling on the shop floor (or even light in some cases), or medical insurance;

and that does not have any identity apart from being an instrument of production. (We will discuss some economic aspects of robot workers in chapter 7.)

Recently, Amazon directed its attention to industrial robots that move finished goods in distribution centers rather than perform factory assembly. Rather than lifting and carrying individual items with arms and graspers, these supply-chain robots locate whole shelving racks of items stocked by humans, which they then move from a storage zone to a picking/packaging station and back again. Robots bring the mobile racks to the workers, who take the appropriate items off and initiate the shipping process. The robots that do this work are not humanoid in the least, sitting low to the ground and looking instead like industrial vacuum cleaners that follow tracks on the floor.

These supply-chain robots build on a long history of material-handling devices known as "automated guided vehicles" (AGVs). These carts or vehicles (either flatbed, enclosed bed, or towing a trailer) can utilize simple navigation methods, such as following magnetic tape on the floor, or more sophisticated ones, such as using lasers, transponders, gyroscopes, and other tools to navigate inside fixed parameters (e.g., inside a warehouse or hospital). The first such system was invented in 1953, and AGVs remain in wide use today.

How Do They Do It? What Challenges Are Being Met?

The generic sense-think-act model tells us little about the complexity and challenges being addressed by robotics.

Structures

Before its sensors, processing power, and actuators can be chosen or mounted, a robot must have a base chassis or other structure. The challenges in this domain are non-trivial. In unmanned aerial vehicles, for example, the robot must be able to fly long distances and provide a stable base for high-resolution cameras, radar, and other sensors as well as weapons. In this and other scenarios, materials science is a critical piece of the equation. Consider the basic physics of many man-made materials: to make a robot twice as tall typically means quadrupling its mass. One humanoid robot, the Willow Garage PR2, stands roughly 5 feet (1.5 meters) tall and weighs some 400 pounds (about 180 kilograms). Such weight introduces multiple issues: the robot's portability is limited, its heavy appendages must be carefully managed in the interest of safety, and its battery life suffers for having to move that much mass. For such robots to gain wider appeal, they must get lighter.

In situations where a robot is going to interact with humans, its structure must convey some sense of familiarity to those who will encounter it. In other words, to

Consider the basic physics of many man-made materials: to make a robot twice as tall typically means quadrupling its mass.

perform its tasks (grasping, detecting, moving), a robot must signal the people around it how to behave. These signals are important to get right; they can be analyzed from many different perspectives: anthropology, semiotics, and psychology, among others. The robot's structure must not only facilitate the functionality needed to carry out its purpose, but also help humans cooperate with it (if only to get out of its way). Whereas some scholars think of a robot as an independent or even autonomous entity, others conceive of it as both aiding and aided by humans. As we implement new robotic equivalents of turn signals on cars or of door handles on buildings, the design choices for robots to be situated among humans will have long-lasting consequences.

Not only must a robot's structure be strong, light, and stable, but its various elements and often the whole robot itself also need to move, adding to the criticality of structural design. Modes of locomotion all have trade-offs: one, two, four, or six legs, wheels, and treads are all options on land. Wheels are extremely efficient but are limited by the smoothness of the terrain. Multilegged locomotion increases a robot's complexity and requires more power than wheels or treads to travel the same distance. Hybrids have also been tried, combining treads and legs, for example.[3] For flight, biologically inspired wings have become feasible, in addition to various types and arrangements of propellers.

Another structural consideration is vibration and damping. For example, to make a robotic surgical tool anchored to the floor that can rise 5 feet, extend 2 feet horizontally, then lower down into a wound site requires a lightness, strength, and vibration resistance not found in most materials. Lightness matters because motors scale up proportionately: a heavier arm requires bigger motors, which both make the device heavier still and limit its battery life, a constant constraint for most autonomous robots.

Very few available components are optimized for robotics application—many parts for robots are still being borrowed from other applications. Whether in the case of motors, actuators, and gears, of microprocessors at all scales, of sensors, of interface devices, or of motive power, robotics stands to gain considerably from advances in some other domain, given the small volumes and high degree of customization currently required in the production of robotics technologies. Partly because important components are often custom made or borrowed from other uses, few robotics efforts have been economically sustainable. Business models are hard to get right, as the exceptional success of the smartphone and video game platforms proves by contrast. Because of its place in a larger economic ecosystem, the Microsoft Kinect sensor bar, for example, is almost certainly sold at a loss, but gives robotics firms and researchers a low-cost, high-performance component well suited to their uses. The haptic (touch-driven) interfaces on the Nintendo Wii provide another example of a mass-market, low-cost tool ideally suited for robotics applications, one that would not have been affordable without game platforms' high volumes.

Navigating trade-offs between engineering, economics, and marketing is difficult. A robotic device can be made to do many things, but deciding in a particular situation what to design in, bolt on, or leave out is hard to get right. Increasing capability by including more degrees of freedom translates into market risk. The history of military drone aircraft provides a relevant example. In 1979, the U.S. Army initiated the Aquila program to build a lightweight reconnaissance drone to radio back images of enemy troop size and location. Additional requirements quickly began to accumulate: night vision, laser target markers, armor against enemy ground fire, secure radio communications, and so on. The drone's weight ballooned, its system complexity escalated, and, of course, its cost soared out of control. What began as a $560 million project intended to build 780 drones ended by spending more than $1 billion on prototypes that didn't work very well nearly a decade later.[4] The iRobot Roomba vacuum cleaner, by contrast, represents admirable control of scope and features in a market-driven scenario.

It is worth mentioning that another new technology related to robotics, additive manufacturing or 3-D printing, can construct honeycomb structures, much like human bones, that combine low resonance, low weight, and high strength. As we will see throughout this book, one constant in the interdisciplinary field of robotics is the cascading effect of improvements in a contributing subfield, whether software engineering, materials science, battery chemistry, or image processing, among others.

Sensors
At the most basic level, a robot needs to have awareness of where it and its attached parts are in physical space. Cameras are one way to achieve this, but they have limitations. First, cameras can encounter lighting situations that diminish their effectiveness: early-morning and late-afternoon sun glare can be blinding; deep darkness is another obvious constraint. Snowfields can reflect significant glare; raindrops can render a lens useless. Visual tricks (such as painting fake potholes on roads) can fool cameras.[5] Turning images into signals can be a challenge for microprocessors and various algorithms: even when a camera captures an image, deriving useful information from that image can be extremely difficult, except when the target is highly constrained, as with license-plate cameras used by police departments and repo men.[6] Once again, advances in one domain frequently lead to advances in seemingly unrelated spheres, and image recognition and processing are key fields for robotics.

Acoustic range finders (sonar and its equivalents) have their uses, within limits: these devices are not especially fast, especially compared with laser range finders such as Lidar. The Global Positioning System (GPS) is useful but not sufficient: it is imprecise for local tasks (such as finding the coffee cup on the table, or the refrigerator in the cafeteria); it can be jammed, and its signal reception can be impaired by human structures such as buildings and bridges. Other proximity sensors, such as bumpers and motion sensors, are also commonly used in robots.

As with electronic networks in a more general sense, a large percentage of a robot's processing power is devoted to monitoring the conditions of the robot itself. Just as a mammal needs to use systemic feedback loops to manage its body temperature or blood sugar levels, so a robot needs to devote resources to monitor and control *its* internal systems. Because few robots can be entirely self-sufficient, one or more radio networks may be in play, connecting the robot to a computational cloud or base station, other robots, external sensors, and the like. Temperatures, power management, system status, and the orientation of various components (at what angle is the left rear leg relative to the body?) all need to be detected. A key example of self-monitoring is a wheel detection system: absent GPS or a local beacon, one of the most difficult pieces of knowledge to provide a robot is where it is, particularly in comparison to 2 seconds or 3 minutes ago. Counting wheel rotations is one basic way to calculate position.

As robots become more advanced, the sensors in their graspers, claws, or "hands" become important. Detecting slippage, for example, must be engineered into an appendage designed both to grasp a slippery beer bottle without dropping it and to squirt ketchup from a plastic bottle without applying too much pressure. Other sensor suites measure radiation (in hazard scenarios), smell (whether of natural gas, explosives, or other materials), and sound, including speech.

In the end, however, collecting sensor data and making decisions based on it remain conceptually and computationally difficult. Thus not only is the environment as a robot senses it often low in resolution or richness, but its sensemaking apparatus is also highly fallible, regardless of the quality of sensor input. If the failure rate of a robot's sensors is sufficiently high, false positives will quickly render a robot experiment useless. Because clean and accurate sensor data can rarely be assumed, error detection and correction are key to improving robot performance.

Computation

Once a robot senses its external and internal conditions, it must first process the sense signals into usable form, so that its control systems can then direct the robot's activities. Although this is not the place to discuss computing architectures, programming languages, or other important topics in computer science and engineering, it may be helpful to touch on a few of the complicating factors that make robotics so challenging.

Time is a tricky business in computer science. Given that very few human phenomena are truly instantaneous, the lag between when a command is given and when it is executed can have important consequences: high-frequency trading on Wall Street is a classic example, where the number of milliseconds of network latency can determine whether an offer is accepted or rejected. Once a robot has to move, time is critical.

The time issue leads to another, related area of difficulty with robots. If you grew up with early computers, you remember the absence of timely feedback: if, after hitting the "enter" key, nothing happened, you most likely retyped the command and hit "enter" again. And, years later, when dealing with an online order website on your much newer computer, if you clicked once and nothing happened, you most likely clicked again, and maybe wound up ordering twelve pairs of socks instead of six. Or maybe neither order registered. When a vehicle is in motion, timely coordination between inputs and actions becomes critical. Given that control systems are less than instantaneous, correcting for various lags between sensor detection, sensor processing, control, and actuation can be difficult. Oscillation will be familiar to anyone who has ridden a bike too fast down a hill: eventually correction cannot be applied with the proper force or speed. One answer is more computing "horsepower," but that generates more heat and requires more power; there are no free computational lunches. More commonly, algorithmic smoothing of both inputs and command variables can reduce the jerkiness and other artifacts of non-real-time processes.

Noise exacts particular costs in a system operating in free space, as opposed to on screen. Given the predictable presence of unexpected sensor inputs, including spurious ones, strict if-then command structures are apt to fail. And given that the robot is operating in and on the physical world, noise and other errors can be self-reinforcing. Fuzzy logic is one approach to noise, and robots often devote a considerable proportion of their processing power to error correction and related tasks.

A major debate in the field of artificial intelligence relates to the noise issue. For decades, it was assumed that a robot needed to use its sensors to first build a map of its environment before interacting with it. With the limited power of the robot's central processing unit (CPU), however, this process took a long time, during which the external environment most likely changed. Thus the robot's cognitive maps consistently lagged reality. Although such a hierarchical approach is required in some complex scenarios, an alternative cognitive architecture has proven to work in certain robotics applications.

Recall that a robot is defined as a machine that can sense, think, and act. In 1986, however, Rodney Brooks, now retired from MIT, proposed that the sense-think-act model could be replaced by a "behavioral" model of "sense/act/sense again/act in light of new information." Rather than act on abstract representations of reality built up from sensing through sensor processing and map building, robots could situate themselves in an environment that they sensed proximately. The outcome of these robots' behaviors seems remarkably "intelligent," in that their actions seem to stem from cognition when in fact they do not.[7]

The emergence of this approach for many robots, including the iRobot Roomba vacuum cleaner, means that efforts can be directed at low-level behaviors: moving, avoiding, reacting in if-then ways. For those who ask if robots are able to be programmed to protect humans, or to seek some manner of goodness (as described by Isaac Asimov in his Three Laws of Robotics), Brooks has had to respond: "They are not." The low-level sense/act/sense again/act in light of new information model produces intelligent-seeming behaviors as an unintentional by-product of multiple small decisions. Robots typically have no "master model" of reality; such models are simply too hard to build.[8]

This is not to say that robots are simply reactive. One major problem area involves path building and planning. If a robot arm with, say, four joints each with x degrees of freedom needs to move from where it is to where a detergent bottle is located, getting its fingers into position to approach the bottle from the correct angle at the correct height without knocking over the potted plant six inches away is no small feat. The robot needs to pay considerable attention to balancing identification of the goal or target (the bottle) with avoidance of obstacles (the plant), within the constraints of its mechanical systems. Some planned routes will be direct but come close to obstacles on the way, so safety concerns (along with a healthy respect for wide error margins in sensor reports) usually emphasize coequal consideration of obstacles and objectives.[9]

In an increasing number of settings, robots, sensors, or both are deployed in groups. The cognitive load on such devices is complicated by the presence of other like-minded actors sharing the same physical space, strengths, limitations, and objectives. Much like birds and insects, swarmbots may not have a "commanding" robot in charge of determining objectives and tactics, but may instead rely on extremely simple rules that produce coordination without hierarchy.[10]

Action

Once sensing and whatever degree of cognition generate a command, robots must execute the command, often in three-dimensional space. This realm differentiates robots from two-dimensional computers in two main ways. First, movement in space is achieved through motors, hydraulics, and other actuators that are neither as precise as pixels on a screen nor situated in as predictable an environment. Second, human-robot interactions occur in time and three physical dimensions, engaging more human sensory, cognitive, and emotional energy than a mouse and keyboard tethered to a desktop. Rules of engagement are thus more complicated. In other words, robots are harder to build in part because humans interact with them in a less constrained manner than with desktop computing.

Apple's Siri voice assistant, IBM's Watson computer, and Google search all provide examples of recent advances in human-computer interface. Rather than simply accepting voice commands that replicate mouse clicks, these natural language systems must learn to understand both multiple voices (unlike earlier systems that were "trained" by a single speaker) and nuances that cannot simply reflect dictionary definitions. Imagine encountering a neighbor's dog on a walk. Saying, "Scratch your back?" "Back off," "Come back here," or "Go back home," can reflect four radically different intentions. Here is where artificial intelligence connects to the state of robotics: when machines interact with people, the people are usually ambiguous about what they desire and how they express those desires. Stating the command versus typing the search term introduces further types of complexity into the interpretive process.

Although the vocabulary of robotic science refers to "control" as a core function of robotic systems, the word "control" is actually problematic. Compared to a human steering a radio-controlled toy race car, for example, different layers of different software architectures may be either more random or less goal oriented than the observed behaviors of the system would suggest. In autonomous robots (as opposed to preprogrammed, anchored factory robots that do the same task repeatedly), a standard architecture has taken shape. At the highest level, human direction is received, and the robotic system plans, sets goals, and possibly changes the shape of the robot. The high-level control hands off to an intermediate layer, where navigation and obstacle avoidance work in parallel. Finally, low-level control over motors and similar devices translates the high-level commands into physical motion, monitoring and minutely adjusting speed, vehicle attitude, and stability. Feedback from lower-level sensors flows upward at the same time that command logic is translated down the stack.[11]

Why So Many Robots All of a Sudden?

Research into robotics has been conducted since the mid-1960s. So why are robots suddenly entering the mainstream in the 2010s? Let us consider the question from the supply-side push as well as the demand-side pull. On the demand side, geopolitics plays a role. Resistance to immigration has intensified the need for automation of mundane tasks for social reasons; indeed, Germany, Japan, and South Korea lead the world in industrial robots per worker. All three countries have both low birthrates and large automotive manufacturing sectors. In the future, personal care robots are expected to help address the consequences of increased longevity along with the desire for low immigration. When robots are used in supply chains and manufacturing, they can bring standardization to precise, repetitive tasks (such as welding or circuit-board assembly) or else free human labor from monotonous, low-value tasks (such as hauling soiled laundry in hospitals).

Although Google, Volvo, Audi, Mercedes-Benz, and other firms are developing self-driving vehicles, many conventional auto manufacturers are introducing robotic or "robot-lite" behavior into their products. Whether in the form of automated parallel parking, path following (lane-departure warning), proximity detection (both in rear cameras for parking maneuvers and front-grill-mounted sensors for anti-tailgating), or GPS, modern automobiles employ a variety of systems whose sensors, logic, and intervention satisfy the basic definition of robot.

As P. W. Singer points out in *Wired for War*, the U.S. investment in robotic battlefield technologies can be tied to the increasing political cost of military casualties after

Vietnam. After the death of more than 58,000 U.S. soldiers in Southeast Asia helped create strong antimilitary sentiment on the home front, high-ranking defense planners along with civilian legislators (most notably, Senator John Warner of Virginia in 2000) began to commit more resources to unmanned systems.[12] The rise in asymmetric warfare tactics in Iraq, Somalia, Afghanistan, and elsewhere further fueled the demand for tools to defend against improvised explosive devices (IEDs) and other weapons of insurgencies. In the longer term, robots in combat could conceivably cut the cost of long-term care for wounded soldiers: the generation of amputees who were injured by IEDs will be expensive to care for over at least the next fifty years, barring breakthroughs in prosthetics, mobility, or tissue regeneration.

Finally, NASA has helped advance the state of robotics, primarily through the development of its Mars landers; indeed, Mars is the only planet in the solar system to be populated solely by robots.

On the supply side, six broad developments combine to make robots more feasible.

1. Moore's Law

For nearly fifty years, Intel cofounder Gordon Moore's observation regarding the number of transistors on an integrated circuit—Moore's law—has held true: transistor density and, with it, overall processing power double roughly every two years. Because many robot tasks are processor intensive (path planning, environmental detection or sensemaking, safety interlocks), increased processor power/speed makes more tasks real time rather than either slow or off board. More onboard computer cores mean more tasks can be attempted or coordinated with a given chip. Advances in graphics processing for video games and in display drivers benefit robotic translation of the real world via sensors to logic that can be incorporated into cognitive processes, and vice versa.

2. Components

The Kinect camera from Microsoft's gaming system (along with the accompanying motion-detection and 3-D software/firmware) means computer vision is more affordable, both financially and computationally. Robotics is borrowing stepper motors from the much larger markets for cameras and car windows, for example. Small, low-power, high-resolution cameras are being made in the millions for mobile phones. Low-power, low-heat microprocessors can also be imported into robotics applications; in addition to chips intended for cell phones, the Arduino microcontroller and Raspberry Pi credit card–size Linux PC bring extremely low priced high performance into laboratories and product-development environments. Mass production of computer tablets has also dropped the price of the touch screen, a previously specialized component.

3. Math

Algorithms for path navigation, image processing, sense-making, and situational awareness can be borrowed from advances in adjacent domains such as search, social network analysis, gaming, video rendering, and natural language processing. Machine learning has emerged as a key field in search and big data analysis to return artificial intelligence research to the forefront after a period of lower interest and funding. A broad open-source movement has extended to robotics to make more code libraries available, so that fewer projects must start from scratch. Rather than writing the world's fifth or ninety-fifth robotic door-opening library, for example, programmers can import or adapt the community's attempt at solving this common problem.

4. Human Talent

Whether from Lego Mindstorms robot competitions, increasing numbers of computer science majors, or the rise of stand-alone departments of robotics in universities around the world, more and better students are entering the field. As industry continues to hire specialists to help build industrial robots, sensor-driven technologies in automobiles and home appliances, and military and aerospace technologies, the appeal of the field combines economic logic with the undisputed "cool" factor of getting to build robots.

5. Money

In the private sector, the dramatic share price increase at Intuitive Surgical along with several high-profile acquisitions of robotics start-ups in 2012 have helped attract venture funding to robotics companies. Google's high-visibility acquisitions of Nest (for $3.2 billion) and Boston Dynamics (for an undisclosed amount) attracted further attention to the field. Not only did Amazon buy Kiva for $775 million, but Japan's SoftBank bought a stake in France's Aldebaran humanoid robotics company for a reported $100 million. Finally, it's impossible to overstate the importance of the massive growth in military robotics spending—for bomb disposal, border surveillance, and drone warfare, to name three current-state projects. Although firm numbers are difficult to derive given the complex and classified nature of defense funding, one industry estimate pegged 2010 defense robot spending at $5.8 billion, projected to increase to $8 billion by 2016.[13]

6. Miscellany

The broad demands of robot building draw on advancements in more developed fields.

• Harmonic drive gears, invented in 1957, are widely used in precision applications such as printing, machine tools, and aerospace along with robotics. Their high torque, compactness and light weight, and ability to achieve much higher ratios than traditional planetary gears in the same physical volume are just some of the many features attractive to robot builders.

•   GPS is ubiquitous, free, and can be used as part of a robot's sensor suite for gross-level position detection.

•   Wi-Fi addresses a key problem in autonomous robotics: how to connect the free-range device to a base station, outboard processor, external camera, or other device. Previous generations of robotics researchers had to adapt complicated and slow wireless protocols or tether robots with cables to build up scaffolding for their research. The availability of cheap, robust wireless networking frees up today's researchers to address more fundamental problems. Wireless networking also paves the way for "cloud robotics," in which heavy processing can be offloaded to servers either off chassis or off premises, with attendant increases in shared learning across devices.[14]

•   Laser scanners came into use in the 1960s, shortly after the invention of the laser. As its cost dropped and reliability improved, Lidar was widely used on autonomous robots (including self-driving cars) for both perception of the environment and object classification.

•   Improvements in software engineering (in debugging, principles of modularity, new types of development frameworks) have helped advance the state of robotics, given that robots require significant amounts of code to run and that there are no dominant robotic operating systems ready out of the box (like Windows). Commercial software such as Mathematica is also useful in sensor processing and other robotic functions.

• Innovations in materials science, whether polymers used to make more lifelike and flexible "skin" in human-facing robots, carbon fiber and aircraft metals used in unmanned craft, or "smart" fabrics that can electrically conduct, resist, and semiconduct as needed, have also helped advance the field. Indeed, through its greatly improved batteries, materials science made possible power innovations in laptops and smartphones, which could never have been funded from the limited research dollars for all of the robotics fields combined.

• Just as computer science took chess as a challenge for its methods for decades, so robotics has taken soccer's World Cup as its ultimate benchmark for team-based independent robots. Begun in 1997, the Robot Soccer World Cup serves as an annual standardized proving ground for research in artificial intelligence and related fields. The official goal is stated as follows: "By the middle of the 21st century, a team of fully autonomous humanoid robot soccer players shall win a soccer game, complying with the official rules of FIFA [International Federation of Association Football], against the winner of the most recent World Cup."[15]

This brief summary shows how broad and deep the interest in robotics has become and only hints at its future potential for many areas of life. It is unlikely that the forces driving advances in robotics—demographics, technological innovations, warfare and politics, ever increasing computing power—will diminish in importance any time soon. The coming decades will provide the proving ground for different types of robots. Which is to say, the future of robotics looks to be extremely bright—and somewhat chaotic.

AUTONOMOUS VEHICLES

In just over 100 years, the automobile and its cousin the truck (both light and heavy) changed the Earth's landscape more than any single technology before theirs. Motorized locomotion in some ways defined the twentieth century, through suburbs, traffic jams, a large percentage of industrialized world's workforce supporting it, and a major role in global carbon dioxide levels. After some major developments such as automatic transmissions and air-conditioning, automotive technology didn't change radically as the world's population has grown from 3 to more than 7 billion since 1960. Thus the impacts of century-old car technology (especially the internal combustion engine) are being felt in India, China, Mexico, Brazil, and elsewhere around the globe: by one estimate, cars and car-related industries generate $2 trillion in annual global revenues.[1]

Many of those impacts are negative, from the environmental costs implied above, to time wasted in traffic, to tens of thousands of traffic deaths every year. To sustain the positive benefits of automobility while mitigating the downsides, there are powerful motives for deploying autonomous or semiautonomous vehicles:

• In dangerous situations, including but not limited to war, natural disaster, or man-made hazards, being able to move supplies into position and people or assets out is highly desirable. The high toll on U.S. supply convoys in deaths or serious injuries from IEDs in the Middle East wars testifies implicitly to the benefits of driverless trucks.

• Nobody can calculate exactly how much time and fuel are wasted in traffic congestion. One 2003 estimate put the numbers at 3.7 billion hours and 2.3 billion gallons of fuel per year. In 2010, another study put the numbers in the same ballpark: 4.8 billion hours of wasted time and 1.9 billion gallons of wasted fuel.[2]

• Cars sit idle most of the time, occupying valuable resources even at rest (as anyone who has paid for parking in a big city can testify). One estimate suggests cars are used less than 4 percent of their lifetime.[3]

• Driving a car safely is challenging. As elders age, reaction times slow, vision can blur, and hearing can weaken. Drunk or inexperienced drivers cause tragedies daily. Ever growing traffic congestion compounds the challenge: patience, skill, and alertness are not always present in the necessary proportions as roads get more clogged every year and commute times lengthen. Globally, the World Health Organization estimates 1.2 million lives are lost in road deaths each year.

How Robotic Vehicles Might Help

Human reaction time and visual calculation are unreliable. It is not hard to conceive of many circumstances under which machines would in fact be better than humans at driving cars. Consider that calculating the speed of an oncoming car is guesswork for a human: in the aggregate, people misjudge their window of opportunity (to make a left turn across oncoming traffic, for example) millions of times every day. Lidar plus processing power makes that calculation trivial for a computer car. Even apart from the Google autonomous car, computers are doing more and more driving every year, whether through traction control or other robotic assists; two speed records for a self-driving car are currently held by Audi,[4] and many other manufacturers are also exploring the possibilities of this technology.

Interlocks to detect a drunken driver, excessive speed, or driving outside a geofenced perimeter are already present. Coupling them to a robotic driver that would take over in a potentially unsafe situation could be popular with some car owners.

Robotic assist need not be all or nothing. Just as traction control keeps cars from spinning out on slippery surfaces, one trend will be ever-greater robotic "help" to human drivers: Tesla launched a program of incremental self-driving capabilities in 2015. That help could also come in the form of Lidar vision (much like night vision has already become an option on some models of automobiles), or faster turning of the steering wheel than a frail driver could muster in a panic situation, or automatic parallel parking, which, again, has already come to market on select models. Wayfinding in the form of automatic GPS and real-time traffic and weather integration could also be appealing: people in rental cars in an unfamiliar city could press the "autopilot" button, perhaps, or commuters heading home at night could be driven along the least congested roads.

Parking spaces in urban areas are scarce and expensive; people pay for parking close to attractive destinations such as shopping, theaters, or sporting events. It's easy to envision the self-driving car as a car-owner's valet, dropping the owner and other passengers at their evening destination, a play, movie, or party, for example, then retreating to some low-cost equivalent of off-airport parking, ready to pick them up again when summoned.

Farther out, imagine cars as automated taxis, a service rather than a product. After directing an automated taxi to drop passenger 1 at that passenger's destination, an algorithm far better than a taxi dispatcher could then route the vehicle to the next closest available passenger, directing it to ferry passenger 2 to *that* passenger's destination, and continue the process. Such a scheme would make traffic lighter (instead of five drivers, let's say, each bringing in a car during the morning commute, one car could carry five passengers sequentially or in a ride share for those who (a) prefer company or (b) want a lower-cost option). Parking spaces could be redeployed to higher purposes. People would have more disposable income: car ownership, counting the cost of gas, maintenance, insurance, and parking, is expensive for an asset that sits idle about 96 percent of its useful lifetime. Unlike a taxi, the robot would never have to go "home," but could stop and refuel or recharge whenever and wherever it was algorithmically optimal. Real-time pricing would balance loads throughout the day: people with time flexibility could pay less for "first available ride after 10:30 a.m.," for example. Uber, a taxicab alternative service built on a smartphone app and drivers who use their personal vehicles rather than medallion cabs, is investing heavily in self-driving cars, including hiring academic talent, and the CEO's public vision aligns with this scenario.[5] In early 2016, Lyft, another car-hire start-up, signed a $500 million deal with General Motors to develop self-driving taxis.[6]

In addition to being convenient, robotic cars driving together can safely travel more closely than those driven by humans. With quicker reaction time and without the human impediments to safe driving (eating, putting on makeup, text messaging, driving impaired, or driving with a physical limitation), robotic cars are more predictable to one another. Roadway capacity and traffic flow efficiency should both increase, making new road expansions less urgent in some cases. On known, premapped roads, robotic vehicles can travel optimally fast: early in its autonomous vehicle research at a test track, BMW programmed a 3 series sedan with the cumulative results of its best test drivers' runs to use as a training tool for young drivers. Although choosing the right lines into and out of a corner, along with the proper shift, acceleration, and braking moves, is reasonably straightforward to automate, BMW's autonomous track car is not yet capable of navigating alongside other vehicles.[7]

How Autonomous Vehicles Are Developing

The past ten years have been a time of rapid advancement in autonomous vehicles. Consider that in their 2004 book

The New Division of Labor, prominent labor economists Frank Levy and Richard Murnane discussed the problems presented by tacit knowledge (things that people know but cannot articulate). They used the example of a delivery truck turning left across traffic as an example of a task that apparently does not lend itself to rules-based definition, in contrast to credit scoring for example.

> The bakery truck driver is processing a constant stream of information from his environment: visual information on traffic light signals, visual and aural information on the trajectories of children, dogs, and other cars, aural information on unseen vehicles (possibly including sirens), tactile information on the performance of the truck's engine, transmission, and brakes. To program this behavior, we could begin with a video camera and other sensors to capture the sensory input. *But executing a left turn across oncoming traffic involves so many factors that it is hard to imagine discovering the set of rules that can replicate the driver's behavior.*[8]

In that same year, the Defense Advanced Projects Research Projects Agency (DARPA) organized a race for autonomous vehicles, with a $1 million prize. The course covered 142 miles of challenging terrain in the Mojave Desert, and fifteen vehicles met the qualifications to start the race after more than 100 teams expressed interest. No human intervention whatsoever was allowed. The most successful entrant covered only seven miles before its rear wheel got hung up and spun until it smoked and possibly caught fire.

In 2005, the agency organized a similar challenge, with a $2 million prize to the first finisher. As DARPA's report to Congress clearly stated, the goals of the race were to

• accelerate autonomous ground vehicle technology development in the [key] areas of sensors, navigation, control algorithms, hardware systems, and systems integration. ...

• demonstrate an autonomous vehicle able to travel over rugged terrain at militarily relevant speeds and distances. ...

• attract and energize a wide community of participants not previously associated with [Department of Defense] programs or projects to bring fresh insights to the autonomous vehicle problem.[9]

Of the 195 teams that applied, 136 submitted the required five-minute video that marked the first project milestone, and DARPA members made 118 site visits, then selected forty semifinalists. After three additional teams were added, the sum total of the team members numbered more than 1,000, many working full-time on the challenge. Semifinalists converged on California Speedway for a qualifying event that simulated some of the features vehicles would encounter on the challenge course. Of the forty-three semifinalist vehicles, twenty-three finished at least one of the three test runs, and five teams completed all three. As it turned out, those five teams were the only teams to complete the final race course, so the qualifying event proved highly predictive.

DARPA's expenditures, including the prize money, totaled less than $10 million, but the return on its investment was extraordinary: the United States moved quickly to the forefront of autonomous vehicle research. More important, global attention is now focused on precisely the areas DARPA identified: sensors, navigation, control algorithms, hardware, and systems integration. An example is both illustrative and instructive.

One team in the 2004 challenge was organized by David Hall, the founder of Velodyne, a Silicon Valley–based company that manufactured subwoofers for home theaters. The team decided not to reenter in 2005, but instead refined and marketed their car's proprietary Lidar system for mapping terrain: GPS is useful for rough positioning but not for identifying traffic lanes or driveways, for instance. At the DARPA Urban Challenge in 2007, seven of eleven finalists had Velodyne Lidar systems on their sensor racks, including the top two teams from Carnegie

Mellon and Stanford (reversing their order of finish at the desert race two years before). Velodyne's unit rotates at up to 900 rpm, generating more than one million distance points per second from sixty-four individual lasers. The 2013 price for the unit was roughly $75,000, a formidable obstacle to commercialization, given that the base Toyota Prius used for the original Google self-driving car was only about one-third the cost of that single sensor. Velodyne Lidar, meanwhile, has become an industry standard, used at Google (where it was called "the heart of the system")[10] and by other research teams. In late 2014, the firm introduced a less expensive 16-laser unit priced at $7,999; in late 2015, Velodyne announced a sub-$500 model would be shipped in the coming year.[11]

The development of self-driving cars is moving along two somewhat opposing paths. At Google, Sebastian Thrun, recruited from Stanford after his team won the DARPA Grand Challenge, helped lead the effort to develop the self-driving car. Not surprisingly, Thrun's philosophy at Stanford—"Treat autonomous navigation as a software problem"[12]—resonated at Google, where tools for handling very large data sets are a way of life. To oversimplify, the Google self-driving car is treated as a software problem, with the car as something of a peripheral to the computers crunching the numbers.

Incidentally, some of the most important numbers turn out to involve not the road or other vehicles, but the car itself, specifically its "pose." Because a 3,000-pound automobile moving through space is subject to the laws of physics, measurements of pitch, yaw, and roll indicate a great deal about where the vehicle is and can go in the immediate future. If the car is perfectly level and has a sensor well off the roof measuring the terrain 50 meters (about 165 feet) ahead, for example, sudden braking will make the nose of the car dive, bringing that 50-meter range considerably closer in as the sensor tilts forward. Thus Google's retrofitted Toyota/Lexus employs a set of sensors devoted to the dynamics of the car itself, along with wheel rotation counters, radar, and Lidar. (The clean-sheet Google car is proceeding along a different track, however.)[13]

Coming from a different direction are the automobile companies—Volkswagen, Mercedes, Volvo, and BMW among them—and their suppliers including Rockwell Collins, Bosch, and Continental. With the incremental addition of more sensors, more processing power, and more actuators every year to the upscale vehicles manufactured by these companies, the robotic nature of these vehicles has increased almost imperceptibly, to the point where *Car and Driver* magazine was led to announce, "The Autonomous Car: You're Already Driving It."[14] Indeed, to achieve fully automated driving may be a small step rather than a giant leap: the sensors for antilock brakes, traction control, safe following distance, lane swerve detection, GPS, parallel parking assist, and rear backup can be further integrated and augmented with more software that can be trained using machine learning techniques. Tesla's autopilot mode adopts this approach and includes no Lidar. For its part, in 2015, Toyota launched a subsidiary to explore both self-driving cars and household assistance robots, particularly for elders. By 2020, its projected investments in this AI venture will total $1 billion.[15]

One experiment in driverless driving has been under way in Europe since 2009. SARTRE (Safe Road Trains for the Environment) exploited many of the technologies just mentioned in a trial that concluded in 2012. A lead vehicle, typically a heavy truck, is driven by a certified human driver. In the original trial, drivers of vehicles equipped with laser and camera systems could signal their desire to join the road train of up to ten cars, called a "platoon," through advance booking, then join the platoon at the appointed time and place. Once in line, each car driver handed off its controls to the platoon. Safe but efficient following distances were maintained while the driver slept, read, texted, or entertained the kids. When the desired destination came close, the driver reclaimed control and exited the platoon. The platoon model has been updated with Wi-Fi interconnection of the vehicles and successful trials were continuing as of 2016. The close following distance enabled by remote control improves fuel economy by 20 to 40 percent. No modifications to the roadways are needed, and the European Union has already reserved a dedicated radio frequency for the system.[16]

Complications

The path to autonomous vehicles will be complex, surprising in its consequences, and some sectors, geographies, and demographics will embrace the technologies faster than others. Although not all of these can be predicted, some of the complicating factors include the following.

1. Law

Changing traffic laws to allow unmanned vehicles on the street might be necessary. Nevada has already led the way, after being lobbied by Google, to legalize self-driving vehicles. Because more requirements and conditions might be added by various interest groups, however, creating a satisfactory statutory environment for autonomous vehicles will be no simple matter.

A key question leaps to people's minds: Who is responsible when autonomous vehicles crash? The answer, it turns out, is tricky.

As Brad Templeton, a Silicon Valley innovator and observer, has noted, when a car accident happens in the current environment, the car owner pays, either directly or indirectly through insurance premiums. With robotic vehicles, liability may well be assumed to reside with companies with deeper pockets: manufacturers, component suppliers, software companies, and the like. Rather than accidents being the result of a moment of bad driving (a personal fact), they will be treated as product liability, a systemic failure to anticipate some eventuality (a corporate failing). But if the judgments against these companies are sizable, it could lead manufacturers to conclude that autonomous cars simply aren't ready for prime time. As Templeton notes, product liability judgments were responsible for the exit of several companies from the small-airplane market: the cost of insurance exceeded the cost of the aircraft since juries assigned blame to the aircraft manufacturer in the majority of cases, even when the pilots were fully responsible.[17]

Templeton also makes a compelling observation about human perception. As behavioral economists including Daniel Kahnemann (a Nobel laureate)[18] and information security guru Bruce Schneier point out, people do a terrible job of rationally assessing risk. Schneier uses the example of sharks: when people see a shark attack on the news, many of them stop going into the water, even though the risk of being attacked by a shark is far lower than that of being bitten by a dog.[19] Meanwhile, cancer, heart disease, and auto accidents kill hundreds of thousands of people a year, but few people give up cigarettes, high-fat diets, or driving to work with the alacrity of the shark response. If 45,000 Americans were killed in traffic accidents, as they were a few years ago, lowering the death toll by 95% might make people *more* distrustful. Probabilistic thinking doesn't match intuition, and fewer but more random deaths feels frightening.

Already many people either refuse to fly or do so with very physical forms of fear, whereas in cars, the illusion of control makes people somehow more comfortable than in a statistically far safer passenger aircraft with a highly trained, tightly licensed pilot at the controls. Robotic vehicles could elicit the same fears, provoking calls for legislation, large jury awards, or both, even if autonomous vehicles were 99 percent safer than nonautonomous ones.

2. Complexity of Environment

In the past, self-driving car schemes have been proposed in which road systems had to have embedded wiring for sensors, dedicated lanes, and other adjustments to an already complicated and expensive infrastructure. Given today's adaptability of self-driving cars to existing roads—albeit heavily premapped existing roads—most roads should eventually be usable by some form or another of autonomous vehicle. But unexpected things happen on the road: deer run out; kids throw water balloons (and worse) off overpasses; plastic bags blow across the road; skateboards skip out into traffic, sometimes with riders still on board;

If 45,000 Americans were killed in traffic accidents, as they were a few years ago, lowering the death toll by 95% might make people *more* distrustful.

and bicycle messengers move very fast and very unpredictably in urban traffic. (Pedestrians and cyclists of all sorts remain difficult to map and react to reliably.)[20] If autonomous vehicles' software says, "Freeze," in every instance of the unexpected, rear-end collisions by human drivers will continue to be the outcome—as they are even in normal driving: from 2009 to 2015, eleven of the fourteen crashes in which human drivers hit Google cars were of the rear-end variety.[21]

Google found early on that driving to the letter of traffic law was not feasible: waiting for adequate space in which to merge led to frustration at the on-ramp; impatient drivers passed on the shoulder. Similarly, in Russia, traffic lane lines are frequently disregarded because of extensive congestion. And what about Los Angeles, Tokyo, or Rome? At construction or accident sites, how will Lidar and computer algorithms deal with flaggers or police officers who sometimes issue verbal instructions? No algorithm fits all driving environments, so how will the self-driving car choose?

Because of the density and labor-intensiveness of the premapped point cloud required by the Google approach, parking lots and structures are not accessible to Google cars. And how do Google cars perceive brake lights and emergency flashers; how do they tell the difference between the light bar on a tow truck and the light bar on an ambulance?[22] Much as in other branches of artificial intelligence, "hard" problems (like map-reading) can be relatively easy, whereas "simple" ones (like how to tell the difference between boulders and pieces of cardboard) can be much harder than anticipated.

Weather can be a significant challenge. Snow obscures lane lines, can present tricky shadows or glare, and affects traction. Rain limits visibility, no matter how good the sensor suite. Road flooding, muddy roads, and sand that drifts over coastal roads can confuse sensors. No amount of predriving and mapping an environment can prepare the autonomous vehicle for every eventuality, so even though being able to ask a human for assistance (via instant video link, for instance) might help in some cases, anomaly detection and processing will likely follow some sort of power law function: 5 percent of circumstances might cause 80 percent of shutdowns, crashes, or other failures.

3. Economics

As we have already seen, early versions of Lidar added $75,000 to the cost of an autonomous vehicle, on top of the cost of radar, wheel sensors, and, of course, computing software and hardware. Moore's law and mass production can help reduce the hardware cost; as software gets better over time, shared libraries of safety-related image processing and related chunks of code may help lower the price for multiple players.

More difficult to anticipate are the subsidies, or lack thereof, for autonomous vehicles. Google might foot part of the cost if drivers will contribute information that aids in ad targeting. Given that Google's most powerful search presence was on the desktop, capturing a share of attention in a car, where people in the United States spend so many hours of their lives, would make business sense for a company that sells ad viewers to advertisers as its core source of revenue.

Once the autonomous technology is proven, insurance companies might make human driving more and more expensive. Statistically riskier drivers—teens and elders—might be required to operate robotic cars in certain situations in order to obtain insurance. On the other hand, states and municipalities might issue tax credits to those purchasing autonomous cars on the logic that such cars, by traveling closer together, can lower infrastructure expenditures and, by having fewer accidents, can lower policing and first-responder expenditures as well. That said, Washington, D.C., to take just one example, issues $80 million in parking tickets per year.[23] What will fill this and other drops in revenue caused by smart vehicles?

And then there are various special interests and constituencies to consider. AARP, already powerful and destined to increase in influence with the graying of the baby boom generation, might welcome autonomous cars as an opportunity to maintain elders' freedom of movement while improving safety. On the other hand, though their influence may not be as powerful as AARP's, people who enjoy driving for recreation are vocal opponents of limitations on their "freedom" as drivers and may see autonomous cars are bringing just such limitations. Insurance companies may well embrace the technology; as autonomous vehicles grow in number, they could make driving so safe that insurance would become much less expensive, with claims and payouts falling off significantly over time. Whereas oil companies are unlikely to support a technology that improves fuel economy, and improves it even more by integration with electric power, faced with stringent miles per gallon targets in the United States, car companies may rush to adopt it, with the proviso that product liability protection is assured.

4. Sentiment

Public opinion is notoriously difficult to predict. It is unclear how robust the long-term markets for autonomous vehicles will be. Whether fear, greed, or novelty wins the day in any given country will have a major role in determining their fate. Part of prevailing sentiment toward the vehicles will derive from the language, the images, the symbols, and the underlying metaphors embedded in the public discussion. "Robocars," "autonomous vehicles," and "self-driving cars" may all refer to the same devices, but the terms and connotations that migrate into wider usage will matter a great deal in determining their acceptance.

Autonomous Vehicle Effects

It's tempting to think of the future of driverless cars as being just like today. But driverless cars could open a host of unintended consequences, some pretty scary (bank robbery escapes) and some deeply profound. Here are several:

1. With driverless cars, transportation as a service rather than an asset could make great strides. Think about what would happen if automobiles became basically land drones, optimizing their path for user convenience (consider a Zipcar-like model), for fuel economy (ride shares), or for off-hours speed of travel. Integrating a tool like Waze, which reports road conditions in real time, along with off-the-shelf optimization software (the same sort of application that helps UPS drivers avoid left turns)[24] and tax or other incentives such as congestion pricing could dramatically change rush-hour traffic flow, insurance rates, and fuel economy. Google is investing in both autonomous vehicles and the Uber car-sharing service. Many things could happen if the two business models converged.

2. Driverless cars could make a significant difference in highway safety and thus public health, assuming even a modest drop in the number of automobile accidents, which currently stands at more than 5 million per year in the United States alone. Driverless cars don't drink and drive, don't text message while behind the wheel, don't fall asleep and drive off the road, and they also don't currently suffer from road rage.

3. Driverless cars could significantly affect how much we spend on automobiles and how we spend it. Think about how much cars cost when they're not in use. A parking spot in an urban metropolitan area can run thousands of dollars per month. Car loans, insurance, and maintenance are huge businesses.[25]

In each of these instances, autonomous vehicles could change people's habits, business profits, government revenues and expenditures, the allocation of public space, and other aspects of civil society.

Now think of how many businesses, activities, jobs, and livelihoods—and how much infrastructure—revolve around the automobile:

• Original equipment manufacturer auto companies: Nissan, Ford, Fiat, and others

• Fast-food restaurants

- Road construction

- Driver's education teachers

- Parking-lot attendants, sweepers, and the like

- Cab drivers

- Toll roads

- Gas stations and convenience stores

- Shopping malls (most have minimal mass-transit access)

- Global auto suppliers such as Michelin, Bosch, Denso, or Delphi

- Car dealerships

- Car washes

- Garages

- Quick oil change franchises

- Auto parts retailers

- Car insurance adjusters, appraisers, claims specialists, and underwriters

- Traffic-related police

- Petroleum drilling, refinement, and distribution

- Corn farming for ethanol

- Bank officers writing and servicing car loans

Now follow the money: who stands to win, and to lose?

Winners

Mapping and sensor companies will provide essential infrastructure for autonomous vehicles. In addition to Google, companies like Bosch, Velodyne, and Continental are making efforts in this market. Audi, BMW, and Daimler joined forces to purchase Nokia's mapping unit in 2015.

If parking spaces could be reduced by a measurable percentage, urban planning could accommodate personal transportation in new ways. Institutions such as hospitals and high schools that spend heavily on parking could find new uses for large pieces of land; indeed, an MIT study found that up to one-third of some cities' land area is devoted to parking.[26] Self-driving cars could also play an important part in many cities' efforts to ban cars from their downtowns: Brussels, Dublin, Helsinki, Madrid, Milan, and Oslo are all moving in this direction.

Intermediaries between passengers and transportation services providers could flourish. If people no longer owned a car for all the hours it sat still, ride-based models like Uber or Lyft could compete with time-share models of the sort used for fractional business jet ownership. At the lower end of the market, Zipcar or Hertz could still be useful providers.

Without hordes of people commuting to work in private cars at the same time, commuters could have more time at home, or more productive time in transit. Even if the timing and duration of the commutes remained roughly the same, being driven should lower blood pressure and increase productivity. Pedestrian accidents should become less frequent.

According to the Centers for Disease Control, nearly 34,000 people in the United States died from motor vehicle accidents in 2013 alone;[27] "motor vehicle traffic" generated 4 million trips to emergency departments in 2010.[28] Any measurable reduction in those numbers would certainly benefit society.

Autonomous vehicles drive better than people, particularly in stop-and-go situations where vehicle-to-vehicle communications or "cloud automobiles" can reduce the accordion-like action of impatient and underinformed human drivers on congested roads. Improved traffic flow through coordination with other vehicles would reduce travel time, increase fuel economy, and reduce net energy consumption. Much like cloud computing, aggregating discrete assets into coordinated use lowers costs and reduces net overhead while dramatically increasing utilization of those assets.

On the downside, in light of the high cost of the failure of autonomous systems, inspecting and certifying them would need to be more stringent, more like inspecting and certifying private planes, and recalls of systems found to be defective might need to more extensive. If the state did not inspect autonomous vehicles directly, it would still need to certify inspection stations.

Much like the countries without copper wire telephone infrastructure that adopted cellular systems more rapidly than the United States, countries that build infrastructure for driverless cars without having to overlay it on traditional roads will be at an advantage.[29]

Losers

With increased asset utilization—cars that are used rather than left to sit idle for those 22 hours each day—car manufacturers might sell fewer units, possibly of different styles (the London taxi might see wider popularity, for example). One potential business model would be to sell transportation as a service: a minivan for hauling kids on vacation, a pickup truck for fetching garden supplies in the spring, a sporty car for the getaway weekend, and an SUV for the ski trip. Rather than owning a car, the customer would pay for access to the right car for the job, by subscription perhaps. One shared vehicle was estimated in 2015 to replace fifteen private ones.[30]

Similarly, parking lots would be less essential. Valets, where they still existed, would be kept on more for pomp than function.

Car dealers and automobile loans might have to specialize more in business-to-business transactions insofar as fleets, or at least pods, of vehicles might be more the norm than personal car ownership. Already, U.S. twenty-somethings buy fewer cars than their predecessors: in only eight years (2001–09), the number of miles driven by U.S. drivers between the ages of 16 and 34 fell 23 percent.[31]

Shopping malls are already in sharp decline. After a prolonged growth phase from 1956 to 2005 when 1,500 malls were built, construction of new malls in the United States has halted. Robin Lewis, who wrote *The New Rules of Retail*, predicts half of remaining facilities will close by 2025, under pressure from Internet shopping.[32] Given the primacy of twentieth-century car culture to the mall, the superstore, and other retail formats, autonomous vehicles would be a key part of the forces reshaping economics and geography in the United States and elsewhere.

Although automobile repair and routine maintenance shops might be busier, given driverless cars' much reduced downtime, it could be that, like cab fleet companies, transportation services companies might manage their own repair shops. And because driverless cars would almost certainly get in fewer accidents, body shops would most likely be quieter. With reduced accident frequency and damage, car insurers should feel quite a pinch from downward pressure on premiums.

Municipalities would see their revenues from speeding tickets, parking violations (and fees), and licensing both drivers and cars much reduced. With fewer drivers, fewer private cars, fewer traffic and parking violations, fewer hours of parking, revenues, police forces, and planning functions for municipalities could all look radically different in fifteen years.

"Drive time" is a crucial time of day for radio advertisers in particular. If texting and video come to play ever larger roles in how people spend their commute time, radio could no longer be the primary driver-friendly entertainment medium. Driverless cars would likely accelerate the move away from traditional radio fueled by satellite radio and digital streaming services' incursion into AM/FM radio broadcasters' territory.[33]

It's noteworthy that Google included a drive-through restaurant in one of its first videos of self-driving car users (the user, Steve Mahan, is legally blind).[34] According to a company spokeswoman cited in the *Boston Globe* in 2009, drive-through customers accounted for between 50 and 60 percent of McDonald's sales.[35] Many car-centric retail formats will need to adjust.

The jobs of taxi and limousine drivers would appear to be endangered in the long run: one estimate put the average cost of a self-driving car ride in New York at 80 cents, compared to $8.00 for a taxi ride. Semitrailer truck drivers, who must navigate busy ports, intermodal facilities, and loading docks, might become more like ferry boat pilots, responsible for only the first and last segments of their journeys.

Thus the question "Do driverless cars—robots of a special type—increase or decrease unemployment?" is impossible to answer with any certainty, given the depth to which the internal combustion, human-operated vehicle is interwoven into the global economy. It's clear there could be whole new sectors born while large numbers of current jobs (those of cabbies, for one example) might disappear. It's also clear that the power of incumbents to affect the process—several states are currently outlawing the Tesla distribution model—will shape the transition as well.[36]

One final category of losers might be unexpected. Because automobile accidents are a prime source of organ and tissue donation,[37] by reducing traffic fatalities, autonomous vehicles could force those in need of transplants to reset their expectations (although some observers predict that 3-D printing of organs and tissues might become a viable alternative as the traditional sources decline).

Trucking

In 2006, the Princeton economist Alan Blinder wrote an influential article on "the next industrial revolution," the one in which service jobs (beyond manufacturing) could be performed offshore. His examples tended toward programming and professional pattern-recognition tasks: equity analysis, accounting, legal research, and radiographic interpretation. In contrast to the blue-collar impact of moving a factory overseas, offshoring services affects individuals at many different income levels, a point Blinder emphasized by pointing to the occupations of nurses' aides and truck drivers as unlikely to get offshored.[38]

The U.S. trucking industry is facing a driver shortage. Even though workers with little education have far fewer options than fifty years ago (when only 10 percent of the workforce had college degrees), truck driving remains a tough sell: isolation, time away from home, and physical ailments from so much sitting and so little healthy food all turn potential applicants away. In addition, stricter enforcement of safety and driving-time requirements add up to more loads than there are drivers. The average age of an over-the-road driver continues to climb (55 as of 2013), and job vacancies mount: 25,000 in the United States, also as of 2013.[39]

Enter driverless vehicles. Although the immediate military advantages of trucks without human operators are obvious given the toll of deaths and serious injuries from IEDs in Iraq and Afghanistan, the civilian advantages would seem to be more long term. Truck driving wages are still relatively low compared to fuel costs and capital investments. Autonomous robotic trucks may well be a decade or two away. Indeed, Mercedes-Benz projects a 2025 rollout for the self-driving tractor-trailer it tested on the autobahn in 2015, pending legal and other approvals.[40] Rather than looking for a "purely" robotic truck, however, once again, there is fertile ground at the intersection of human and compu-robotic capability.

Questions We Need to Address

Who's Responsible?

The notion of vehicular autonomy requires careful thought. Whether in the service of a military commander or a civilian doing household errands, robotic vehicles are always doing someone's bidding; the autonomy is relative, not that of a teenager being given the keys to the car. The Google self-driving car cannot decide whether to go to the ice cream stand, the movies, or the mall. Once given a destination, obviously these vehicles can optimize routing, estimate travel time, reroute in the event of heavy traffic, and do many other useful things.

Thus one question bears directly on the relationship between users and the autonomous vehicles they are using or own. When a vehicle causes harm or disruption, who is responsible, especially in cases in which nobody is visibly in control of it? A RAND study reinforced the importance of the role of liability concerns in slowing adoption of these technologies.[41]

Where's the Money?

What will business models for autonomous passenger vehicles look like? In the taxi or Uber examples, people pay a service to take them to a destination, and it's easy to imagine extensions of those businesses, albeit without drivers to pay. Google, for its part, has a multibillion-dollar content navigation business already in place. Will Google car riders agree to trade watching ads for taxi fare? What entities will be more aggressively or most favorably positioned to become content intermediaries for the time freed up from driving ?

It's easy to imagine the manufacture of autonomous passenger vehicles as a continuation of traditional auto manufacturing. Indeed, many traditional automakers are experimenting with further advancement and integration of robotic technologies already available (driver-warning systems, antilock brakes, automated parallel parking). But what happens when the passengers' attention is the relevant asset? Comcast bought NBC Universal to own content to push over the company's cables. Might Sony, for example, integrate home entertainment into a self-driving car? What about Samsung or Microsoft? Apple has plans to integrate traditional cars as i-device peripherals. In the end, the autonomous passenger vehicle business model could be closer to that of television, smartphones, or tablets.

What Happens When Things Go Wrong?

Autonomous vehicles run on software, and software is never perfect. What happens when weather or road conditions (fog, rain, snow, flooding, or sinkholes), road construction, or chaotic human driver behavior confuses the guidance systems of these vehicles? How much user input will be available, whether to take control in an emergency, reboot the systems, or even push? Who will refuel autonomous gasoline vehicles in the forty-eight states without gas station attendants?

How Do We Get There from Here?

Path dependence is a powerful force; early design decisions will shape future innovations. Will the Google approach (attaching motive hardware to a big data–processing platform) lead the way, or will traditional manufacturers incrementally add new sensors and more processing to extrapolate from the present state? Who will be the guinea pigs among state licensing and regulatory agencies, insurers, passengers, and dealers?

How Will Incumbents Fight Innovation?

Uber has had to battle taxi and limousine commissions in city after city. Airbnb has been sued by the state of New York. The record industry lobby sued music downloaders. General Motors bought, then dismantled streetcar lines to discourage mass transit. Oil companies have lobbied against subsidies to alternative fuels. In the coming battle, given such enormous financial stakes and such longstanding business practices, the vested interests will not submit quietly.

What Role Will Geography Play?

It's difficult to see that any one place will have a natural advantage in rapid adoption of self-driving cars. Certainly, the more challenging the traffic environment, the tougher the programming task. But given moderately adequate infrastructure, good cellular coverage, necessary levels of wealth and investment, and the modest rule of law, plenty of countries could help launch the technology. National champions—companies with high levels of national prestige and home-government support such as Citroën, Michelin, Continental, Bosch, or Fiat—could spearhead wide-ranging development programs in their home countries.

What Will It Cost?

For all the attractive savings in commute time and fuel consumption, quite apart from the significant gains in rider safety, what will it cost the person riding to work, the airport, or a night out? Currently, economies of scale have not kicked in for the key sensors, the computing platform is experimental, and investments in the new mode of operation (as in sensors) have not fed back to cost savings in the traditional platform (in lighter bumpers, for example). Indeed, the cost of its robotic systems currently far exceeds the cost of the autonomous base vehicle, probably by a factor of 2. How long will it be until a first-time car buyer can consider buying a self-driving car on a competitive cost basis? Reaching this breakeven point will, of course, require considerable innovation at the business model level: it's unlikely that an autonomous car would be a direct competitor to an entry-level traditional car without some package of included (read "subsidized") services.

Conclusion

The biggest question regarding autonomous vehicles relates to our ability to think beyond current limitations, costs, and habits. Who will have the freedom to think from a fresh perspective and completely reinvent personal mobility? To use a computer term of the 1990s, the autonomous vehicle platform is in search of its "killer app," its breakthrough way of framing and addressing the need for an alternative form of mobility. The limits of the technology are falling away far faster than are existing assumptions and stereotypes; when will the reverse be true?[42]

WARFARE

The role of robotics in modern warfare is changing rapidly, and with the most strategic of implications: how war is waged, where combat is occurring, and the risks that apply to combatants on both sides are all being redefined. Moral issues of warfare, never simple to navigate, are also becoming more complicated.

Motivation

As its broad mandate, DARPA, the U.S. Department of Defense's R&D organization, strives "to make pivotal investments in breakthrough technologies for national security."[1] The vision for DARPA's Tactical Technology office, where robotics research is primarily centered, is to "rapidly develop new prototype military capabilities that create an asymmetric technological advantage and provide U.S. forces with decisive superiority and the ability to overwhelm our opponents."[2] Ronald Arkin, a Georgia Tech robotics researcher, asserts that the TTO's vision fuels four interrelated objectives:

• *Force multiplication*—where fewer soldiers are needed for a given mission, and where an individual soldier can now do the job of what took many before.

• *Expand the battlespace*—where combat can be conducted over larger areas than was previously possible.

• *Extending the warfighter's reach*—to allow an individual soldier to act deeper into the battlespace; for example, seeing farther or striking farther.

• *Casualty reduction*—removing soldiers from the most dangerous and life-threatening missions.[3]

Several of the concepts involved in achieving these objectives deserve deeper consideration. First, "asymmetric warfare" characterizes the U.S. wars of the twenty-first century. Two contrastingly equipped and motivated sides fight, with each side seeking to take advantage of its superior assets: the United States deploys massive technological superiority, whereas the insurgents capitalize on their greater ideological appeal in a native population. That the insurgents' ideology might sanction suicide bombers, for example, or human shields in the form of schoolchildren or hospitals renders some of the United States' technological advantage moot. At the same time, the dearth of Arabic speakers in particular, and of cultural sensitivity and understanding more generally, within U.S. forces makes capturing the "hearts and minds" of affected populations highly problematic. As opposed to air war with Germany or naval war with Japan in World War II, when the two sides were much more closely matched in both doctrine and arms, present-day asymmetric warfare drives a search for not merely better equipment—such as the jet engine in World War II—but entirely new modes of combat.

Hence the concept of "expanding the battlespace." The theoretical construct known as a "three-block war" was introduced in the 1990s to describe the potential for an army or marine corps to be fighting armed conflict in one area, performing peacekeeping nearby, and delivering humanitarian aid in the third "block" of a hypothetical town. Although it cannot serve as a literal strategy and does not include the essential tasks of nation building that were undertaken in Iraq, the complexity of modern war makes application of classical military doctrines problematic. When the field of battle is no longer defined by "sides" that "capture" territory, the role of the armed forces, especially on the ground, can change dramatically.

As far as "force multiplication" goes, the composition of a modern army or navy is considerably different from what it was two generations ago. Public opposition to bloodshed has changed the politics of military budgeting and deployment. The military's tasks (establishing the rule of law vs. clearing mines), approaches (frontal warfare vs. counterinsurgency), motivations (protecting shipping lanes vs. a "war on terror"), and combat personnel (draftees vs. volunteers including women and people of multiple sexual orientations)—all are different in 2016 from what they were in 1976.

Thus a wide variety of factors helps motivate the development of robotic technologies for warfare, and the considerable impact of billions of defense-related dollars on the state of robotic art and science means that many nondefense developments likely owe their existence to one of these battlefield imperatives. For these and other reasons, a closer look at military robots is essential for an adequate understanding of robotics more generally.

Types and Forms of Military Robots

The range of applications of robotics to military use is broad and increasing every year. The following is intended not as a catalog of the robotic arsenal, but rather as an introduction to basic types and forms of military robots.

Air

To date, unmanned aerial vehicles (UAVs) have been primarily used for reconnaissance purposes, although Predators and the much more heavily armed Reapers have also been used to deliver ordinance. UAVs range in size from a few pounds to substantial: compare a hand-launched Raven, at 3 feet in length, to the Global Hawk, at 44 feet (about the size of a corporate jet) and 26,750 pounds (about 13.5 tons). A constant dilemma lies in choices of instrumentation: keeping a UAV light and to an extent "dispensable" is achieved by minimizing the sensor load. On the other hand, single-purpose aircraft are more difficult to maintain at a suitable level of readiness, especially when procured years before their current deployment, which might entail the need for new sensors. Thus many UAVs have suffered from "requirements creep" and become heavier than planned, at the expense of performance, time aloft, and budget projections.

As of a 2012 report by the Congressional Budget Office, the U.S. military owned 10,767 manned and roughly 7,500 unmanned aircraft. The vast majority (5,300) of those UAVs are U.S. Army Ravens, 4-pound reconnaissance aircraft launched by throwing them, much as you would toy gliders. The same report notes that, whereas projected total spending from 2001 to 2013 for unmanned aerial systems (UAS, which include "ground control stations and data links" needed to operate UAVs) is more than $26 billion, manned aircraft still got 92 percent of Pentagon funds for aircraft procurement.[4] In 2009, the budget for one F-22 manned fighter could have bought eighty-five Predator drones, for which training and operating costs would have been significantly lower as well.[5]

The UAVs that have seen the most service are listed in table 6.1.

Table 6.1 Specifications for most commonly deployed UAVs as of 2012

Name	Function	Length	Wingspan	Time aloft	Ceiling
Global Hawk	Surveillance	47.6 feet	131 feet	28 hours	60,000 feet
Predator	Offense; surveillance	27 feet	49–55 feet	24 hours	25,000 feet
Fire Scout (unmanned helicopter)	Targeting; surveillance; fire support; reconnaissance; situational awareness	24 feet	27.5-foot rotor diameter	8 hours maximum; 5 hours fully loaded	20,000 feet
Raven (hand launched)	Situational awareness	3 feet	4.5 feet	60–90 minutes	n/a

Source: Summarized from Congressional Research Service, "U.S. Unmanned Aerial Systems," January 3, 2012, http://www.fas.org/sgp/crs/natsec/R42136.pdf.

The capabilities of drones compared to manned aircraft and satellites can be summarized as follows:

1.   Drones can linger over geographies of current interest for 24 to 48 hours at a time, at altitudes above the limits of enemy fire and at no risk to human pilots, and deliver high-resolution, real-time imagery, whereas satellites can only fly over these areas for brief times, do not deliver imaging and other intelligence in sufficiently high resolution or in real time, and must be tasked long in advance. The drones' ability to stay aloft for long periods has numerous benefits. Large physical areas can be surveyed, smaller ones studied in greater detail, and targets of interest tracked. One challenge going forward, as with other forms of video surveillance, is to automate the processing of tens of thousands of hours of imagery.

2.   Unlike manned fighter aircraft, drones can be launched from and return to airstrips located far from their targets. As P. W. Singer put it in his book *Wired for War,* a "Global Hawk can fly from San Francisco, spend a day hunting for any terrorists in the entire state of Maine, and then fly back to the West Coast."[6] Pilots are kept out of enemy airspace, avoiding the strategic and diplomatic risks entailed therein.

3.   Drones keep human pilots out of harm's way, deliver many of the same air-to-ground capabilities that manned

aircraft do, and because they are mechanically simpler, spend a higher proportion of their time either in flight or ready to fly. The F-22 fighter jet, for example, was reported by the *Washington Post* in 2009 to require thirty hours of maintenance per flight hour,[7] whereas the Predator drone was reported to require only one hour of maintenance per flight hour in a border patrol scenario (combat readiness figures were not available).[8] Indeed, drones provide far better bang for the buck in both procurement and operational costs; in a time of shrinking defense budgets, manned fighter jets that can claim dogfight superiority are an expensive proposition, especially given how little air-to-air combat has occurred in the past forty years.

4. Training drone pilots takes far less time at far less cost, given the extensive, and expensive, air time required of flight school pilots. Drones can tirelessly perform mundane tasks, such as flying in a routine pattern to cover a large area (much like a farmer plowing a field), that human pilots would find boring and likely perform with less precision. Because inanimate objects can withstand greater G-forces than human pilots can, drones should eventually be able to outmaneuver manned craft. And drones can fly into dangerous conditions (radiation, volcanic activity, enemy fire) without risking the life or health of human pilots. Indeed, sacrificing a drone to get enemy antiaircraft batteries to turn on their tracking systems so they can be

located would seem to be an entirely justified strategic trade-off in certain circumstances.

5. Drones' lighter weight and smaller engines make them easier to ship into service, harder to detect, and cheaper. Smaller engines mean less noise, less pollution, less fuel usage, and cheaper, lower-spec components. Keeping a pilot out of the cockpit of an aircraft simplifies design, lowers weight, and maximizes use of all its available resources (including fuel) to deliver either surveillance or weapons.

Sea

To date, seawater presents a more challenging environment than air or land for unmanned vehicles, so there is less news to report. Saltwater is highly corrosive; wind, tides, and currents make navigation less amenable to automation; fog and rain stress sensors for robots in all environments; waves can present stressful physical environments for sensitive electronics; and secure radio access to offshore and especially underwater craft is limited.[9] Nevertheless, unmanned vessels' potential for dull tasks such as monitoring or mapping the ocean floor or dangerous work such as mine detection and detonation is sufficiently promising that several initiatives are under way. To date, there are fewer production drone craft in the Navy than in the other armed forces for a number of reasons, some technical, some historical, and others organizational.

Unmanned underwater vehicles (UUVs, also called "autonomous underwater vehicles," AUVs) must operate independently (given the limits of radio signals), must have a power source suitable for low-noise unmanned operation, and must be protected from being captured and potentially repurposed by an enemy. Thus armed craft are a longer-term objective, but surveillance and mine clearing fall within a near-term time horizon. The Remus is a converted torpedo made by a Norwegian firm, for example, and there are reports of unmanned minisubs launched from conventional submarines. The Seaglider, made by the same firm that builds the Remus, is powered not by an electric motor but by small changes in buoyancy, allowing it to collect data while at sea for months at a time, offloading readings from the surface to satellites.[10] As elsewhere, the line between scientific data collection and military applications can be blurry at times, but, to date, many UUVs have been used to carry out research.

A new UUV is in the midst of sea trials. Weighing 6,200 pounds (3.1 tons), the Proteus is 25 feet long and can be operated in either autonomous or manned mode. Payloads vary, but SEALs, bomblets, and other cargo can be accommodated. The Proteus has a theoretical range of 900 miles on a battery charge, at a top speed of 10 knots (11.5 miles per hour), and can descend to roughly 100 feet.[11]

An unmanned surface vehicle (USV), the Spartan Scout is a 36-foot boat capable of speeds of 50 miles per hour. It is equipped with a wide array of sensors, making it more suited to reconnaissance than to the mine-sweeping work done by the Remus, although it carries a fifty-caliber machine gun aboard, along with speakers and microphones for remote interrogation of suspicious surface craft. The Spartan Scout was used in the Persian Gulf during the Iraq war in 2003.[12] The Israeli navy claims to have deployed the first USV, the Protector, a rigid inflatable craft 30 feet long (there is now a 36-foot version), for surveillance and reconnaissance.[13]

Land
Robotic ground systems have advanced rapidly in the past fifteen years, with different "species" emerging. One way to differentiate among the unmanned ground vehicle (UGV) variants is to look at their mode of locomotion: UGVs that use wheels or treads constitute one class; those which use legs, another.

Wheeled   The world's largest robotic vehicle is a 700-ton dump truck not currently used in combat but in mining. The far smaller Oshkosh TerraMax is in fact sizable by most nonmining standards and used by the military for supply and reconnaissance. Weight and other specifications are not available, but its "drive by wire" technology, already used for many systems (throttle, brake, etc.), can be readily adapted to remote control or even autonomous operation.

At the other extreme of UGV size, the Boston Dynamics Sand Flea weighs 11 pounds and has a gas piston for launching itself 30 feet into the air and onto a roof. Gyro stabilizers keep the device oriented and level during flight so as to maintain usable video imagery. Although current versions provide eyes and ears, it's not hard to imagine a "smart grenade" on a similar platform.

Tracked   Tracked UGVs have come to the forefront in the Middle East wars since 2001. Most are produced by two Boston firms spun out of MIT. One firm, iRobot, makes the PackBot (see figure 6.1), a 24-pound remote-controlled tracked vehicle that has been used to discover and disarm thousands of improvised explosive devices. Building on a reconnaissance device, later generations of PackBots and similar vehicles have added mechanical arms and grippers to move small objects and disarm or safely detonate explosive devices. Other models have added a variety of sensor, camera, and software configurations capable of localizing gunshots (to help spot snipers); detecting hazardous materials, toxic gases, and radiation; and recognizing faces. As of 2013, 2,000 such devices were deployed in Iraq and Afghanistan.[14]

A second MIT spinout company, substantially older than iRobot, Foster-Miller also makes tracked combat UGVs, now under the name "QinetiQ," the company that

Figure 6.1   iRobot PackBot. Photo courtesy of iRobot.

acquired it. The TALON is bigger, at about 125 pounds (depending on configuration), faster, and comes in more configurations than the PackBot. Most notably, the SWORDS (Special Weapons Observation Reconnaissance Detection System) version of the TALON is loaded with a choice of weapons (see figure 6.2), although its deployment has been limited.[15] Available armament options include a rifle, shotgun, machine gun, and grenade launcher.

Figure 6.2 QinetiQ SWORDS (weaponized TALON robot). Photo: U.S. Army.

Although the SWORDS is believed to be the first armed UGV to see combat, more common scenarios see the tracked combat robots run into buildings thought to be booby trapped, into caves, or around corners before human soldiers expose themselves to risk. At times, the robots have carried a grenade or similar cargo into an adversary's territory, sometimes saving lives by sacrificing themselves.

More tracked variations are emerging. The iRobot FirstLook weighs only 5 pounds and is designed to be thrown through windows or otherwise deployed as soldiers' eyes and ears; the ReconRobotics Scout, which weighs just over 1 pound and whose heavy metal wheels can break glass, can safely be thrown more than 100 feet. The Scout automatically rights itself upon landing and, like the FirstLook, transmits video feeds over a wireless network; in February 2012, the U.S. Army ordered 1,100 of these "throwbots."[16] Like other reconnaissance robots, the Scout is also being rapidly integrated into police, fire, SWAT, and other domestic public safety applications after having proved itself in combat.

Legged   Though still an emerging class of ground robots, legged UGVs may well represent the future of combat robots. DARPA, the U.S. Department of Defense's think tank, has been pushing their development, in part through its support of Boston Dynamics, yet another MIT spinout firm. The firm's publicity videos are impressive: its Cheetah robot has exceeded 29 miles per hour on a treadmill; its LS3 robot, essentially a pack mule, has demonstrated a striking capability to traverse uneven ground, although it is as loud as a lawn mower.

In 2015, Boston Dynamics introduced a new version of a humanoid robot (Atlas II) designed to be used a DARPA challenge to cope with a reactor disaster scenario like Fukushima. The competing robots had to climb through wreckage, be able to withstand heat, radiation, and other adverse conditions, and manipulate valves, levers and similar controls characteristic of a nuclear reactor. In navigating an environment designed for humans, they also had to be able to open various kinds of door handles, use tools, and perform complex tasks.[17] For some teams participating in the DARPA challenge, the Atlas robot and its predecessor (see figure 6.3) served as a shared platform to build upon if they lacked resources to build their own hardware.

Other robots that mimic biological structures are also in development: crawling legged robots are designed to meet the new requirement for greater flexibility and mobility in combat scenarios. Thus the Boston Dynamics RHex (see figure 6.4), which moves on six unhinged "legs" that are essentially rigid arcs of tread material and weighs about 30 pounds, can traverse extremely challenging terrain. Indeed, the RHex can even negotiate ditches and culverts, where insurgents began hiding IEDs once they discovered tracked robots robots could not fit into those spaces.[18]

Autonomy

To date, the large drones that get most of the public's attention are remotely piloted by humans via video links. Smaller, autonomous drones like the Raven can fly to GPS waypoints and can land on their own, but they carry only

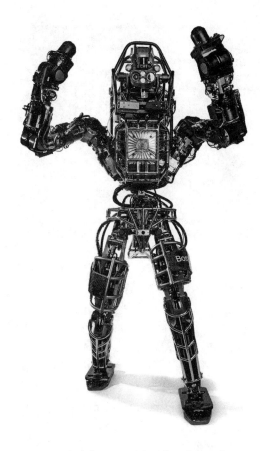

Figure 6.3 Initial version of the Atlas robot used in DARPA competitions.
Photo: Boston Dynamics.

Figure 6.4 RHex surveillance robot. Photo: Boston Dynamics.

sensors. The U.S. Navy has developed a next-generation semiautonomous UAV (the X-47B) that has successfully landed on an aircraft carrier, one of the hardest tasks in aviation. Beyond takeoff and landing, however, a debate over autonomous targeting and firing of weapons has already begun. The Phalanx ship-borne antimissile system is essentially a robot that detects close-in oncoming missiles and fires a Gatling gun at 4,500 rounds per minute; humans are only tangentially in the loop given the imminent nature of the threat. (Because of their distinctive shape, Phalanxes are known in the U.S. Navy as "R2D2s" and in the British navy as "Daleks.") There are deep concerns, however, about autonomous robots firing on people, whether from an airborne drone or a land-based platform.

Three basic relationships between human operators and lethal robotic delivery systems can be considered. With "human-in-the-loop" configurations, a human sees the information provided by the UAV or other device and if the firing criteria are met, issues a direct order to fire whatever weapon is available. One step removed is the "human-on-the-loop" weapon, in which a human operator can override the robotic decision after the information has been processed and a target identified autonomously. In the future, "human-out-of-the-loop" architectures would allow a robotic weapon to detect, select, and fire on a target without human intervention.

The advantages of a fully autonomous weapon are many. In tense border or other scenarios, such weapons minimize the chances that the human operators will be killed or incapacitated in a sneak attack; at the North Korea–South Korea border, semiautonomous lethal robots built by a division of Samsung are already deployed. Robots do not fall asleep, or object to harsh conditions, or "go native."

Georgia Tech roboticist Ronald Arkin is helping to develop autonomous robots for the U.S. Army, Navy, and other agencies. He argues that robots can *potentially* be better soldiers than humans for six reasons, which I summarize here:

1.   Because robots do not need to "kill or be killed," their algorithms can be more conservative than those taught human soldiers by boot camp and field experience. That is, a robotic warrior can be programmed to sacrifice itself in ways that are morally and operationally difficult for human soldiers to do so.

2.   Eventually the robotic sensor suite will be better (more robust, more extensive, more redundant, more integrated with sensor networks) than the sense organs of a scared, often confused human.

3.   Robots lack emotions, so retribution, fear, and hysteria can be removed from the targeting equation in ways they cannot with human soldiers.

4.   Behavioral psychology powerfully illustrates the framing biases of humans: we see what we want to see, or fear seeing, even if it is not there. Robots lack these biases.

5.   Sensor integration through greater processing power, better algorithms, and lower probability of information overload gives the robotic warrior an advantage over the human.

6.   Robots can be impartial. If they act as observers and recorders of activity within a mixed human-robot team, the robotic elements can act as a check on human ethical infractions and other breaches of discipline.[19]

To his credit, Arkin is also frank about the potential for ethical failures of robotic warfare. Technical, political, operational, and human rights issues are only beginning to be explored. Concerned organizations such as Human Rights Watch[20] and the United Nations[21] have raised cogent objections.[22]

• Who will assess the danger posed by robotic warfare to civilian populations, with human judgment invariably clouded by bad or missing information, deception, or mistaken narrative framing?

• How will the limits of AI rules engines be recognized and respected? In many Arab countries, guns are sometimes fired toward the sky at celebrations like weddings. It is not difficult to imagine an autonomous aircraft spotting gunfire from an AK-47 and retaliating, for example. Although this example is hypothetical, another is not: in 2008, a U.S. air strike was reported to have killed forty-seven members of a wedding party escorting the bride to her groom's house in Afghanistan, including thirty-nine women and children. An Afghan inquiry claimed none of the dead were connected to al-Qaeda or the Taliban.[23] No U.S. inquiry findings were ever made public.

• Will a robot be able to recognize and act on signals of surrender? And will vanquished fighters surrender to machines, an act that in some cultures would reflect cowardice?

• Who will be responsible for a robotic kill gone wrong? The robot's manufacturer? The robot computer's programmer(s)? The pilot or video analyst operating or overseeing the robot, who could be a contractor and not a combatant?[24] The presiding officer? The commander-in-chief?

• What will happen if going to war becomes too easy, given the lower probability of human military casualties?

• What will happen if going to war or fighting a war involves micro-second decisions, much like those in Wall Street's high-frequency trading? If algorithms are fighting algorithms, first-mover advantage could be substantial.

• If warring nations can kill enemy warfighters with little or no danger to their own, will such wars be viewed as unjust?[25]

• Will the technical problems of facial recognition, lethal and nonlethal weapons (such as rubber bullets or sonic weapons), potential radio interference, false positives and false negatives ever be satisfactorily solved?

• What happens when the robot refuses a human order or override, or warfighters hesitate to second-guess a computer? In 1988, an early automated guided missile system called "Aegis" on the USS *Vincennes*, a cruiser stationed off the coast of Iran, targeted a civilian airliner and blew up 290 passengers—even with a human in the loop. Given limited information and international tensions in the area, it is not unreasonable to believe the operator could have been worried about failing to defend against an enemy attack had he not allowed the system to fire.

• What will happen if overrides are made or refused for political, psychosocial, or other unpredictable reasons? What will happen in a *Dr. Strangelove* scenario?[26]

• How will the vanquished react to losing to a force of robots?

• What will happen when (not if) combat robots are co-opted? The video feed from U.S. Predator drones was for a time only lightly encrypted and thus easily viewed by the enemy. Hidden software has already been discovered to be embedded in microprocessors,[27] so there is every reason to believe robots could be hacked to the home country's disadvantage, either after capture or through sabotage.

Consequences
The use of unmanned vehicles in warfare has had many unexpected consequences. The following examples only illustrate the breadth of issues raised so far; the future will grow more complicated still.

• After the Vietnam War became known as "the living room war," shaped in large measure by film aired on the TV networks' evening newscasts, and Gulf War I, as the "Nintendo war," based on the night-vision footage of missiles and smart bombs vividly blasting Iraqi targets, more recent warfare in Iraq and Afghanistan have contributed drone-captured YouTube videos to the public's viewing menu. Unlike the unsettling footage of U.S. troops in Vietnam that helped turn public opinion against Lyndon Johnson and later Richard Nixon, "war porn" from Iraq and Afghanistan is pro-U.S., often circulated by the U.S. military, and can draw more than a million views per clip.[28]

• Pilots of unmanned vehicles are physically safe, sitting in a cubicle thousands of miles from either the drone's target or sources of enemy fire. The emotional toll, however, in only beginning to be understood.[29] Fighting a remote war for twelve hours, then returning to a suburban life, with a family and the expectations thereof, is more than jarring. The lack of "band of brothers" unit cohesion among drone pilots is another factor: sharing hardship would build a camaraderie that could help the pilots process the intense emotions of their job. Witnessing a U.S. unit being ambushed and being powerless to help is reported to be a particularly wrenching experience.[30]

• The tactical and strategic advantages provided by UAVs can also convey unintended cultural meanings. U.S.

enemies may well view sending unmanned vehicles to do the killing, with no risk to U.S. pilots, not simply as technological cleverness but also as cowardice. In the words of Indian Muslim author Mubashar Jawed Akbar: "In war terms, if you are not willing to sacrifice blood, you are essentially a coward."[31] Thus the technology that keeps U.S. warfighters out of harm's way may also be motivating its enemies to greater resistance and legitimizing new anti-U.S. beliefs and behaviors.

• For anyone who has been upset by an interrupted or lost cell phone connection, understanding the radio frequency (RF) landscape in a military engagement is both a sobering and a daunting challenge. Given how much radio energy is being produced and consumed—in the form of encrypted communications, GPS, video feeds, reconnaissance by multiple technologies, radar, and attempted or actual jamming of all of the above—it is not surprising that there can be serious disruptions.[32] What works well or passably well in controlled trials can be difficult or even impossible in the data smog of battle, especially given the lack of shared infrastructure and the need for encryption, which multiplies the size of even simple messages and amplifies demands on the infrastructure. One underappreciated aspect of the future of robotic warfare will be the innovations and counterinnovations with regard to wireless control, oversight, and jamming of autonomous devices.

One underappreciated aspect of the future of robotic warfare will be the innovations and counterinnovations with regard to wireless control, oversight, and jamming of autonomous devices.

Conclusion

Throughout human history, war and conflict have given rise to major technological advancements. Gunpowder, steamships, aircraft, nuclear power, GPS, and the Internet are just a few of such innovations. Like other military devices before them, unmanned vehicles have great potential for both good and bad. Drone aircraft could transform humanitarian relief, for example; legged ground robots could become superhuman firefighters or disaster rescuers. Just as likely, however, killer robots could do the bidding of drug lords, religious extremists, or alienated youths. Figuring out how to frame the social discussions that are emerging from robotics labs and the mass production of robots is a high-priority task for politicians, legislators, judges and juries, and citizens the world over.

ROBOTS AND ECONOMICS

Given that robots have been performing assembly-line tasks for nearly fifty years, it is somewhat surprising how little is known about the relationship of robots to productivity and employment. As robotics expands from assembly-line to supply-chain and eventually to service applications, the use of robots will affect more people, and thus presumably move the issue closer to mainstream discussion.

Do Robots Take Human Workers' Jobs?

There's a thought experiment in which you're asked to take the position of an engineer in 1890 and project the amount of horse manure in New York City in 1920. Linear extrapolation produces a frightening result, which, of course, never happened: the invention of the automobile shifted the externalities of transportation, and, instead of staggering quantities of horse manure in 1920, we got suburbs, McDonald's, high-speed roadways, and dozens of other side effects, starting in 1930 and continuing on through the present.

Something of the same situation applies today with regard to the impact of information, and soon robotic, technologies on unemployment. It's easy to assume that ATMs put bank tellers out of work, for example; President Obama implied as much in a speech in 2011. The evidence, however, suggests otherwise: the number of bank tellers increased from about 450,000 to 527,000 in the first twenty years of ATM technology. Whether there might have been an even larger job increase without ATM technology is, of course, impossible to know.[1] The same goes for robots and automobile jobs. Unemployment in Detroit has a variety of causes, and it's impossible to isolate robots as a decisive factor, given the rise of Japanese and later Korean automakers, ongoing subsidies to "national champion" automakers in some countries that prevent a market shakeout, declining rates of automobile ownership, the state of labor unions outside the industrial Midwest, and the contribution of pensions and health care costs to Big Three labor economics.

Note that both self-serve gas pumps and ATMs are early examples of human-robot partnership. Self-service—whether at Amazon.com, the gas station, the airport kiosk, self-checkout, or wherever customers often do the work of employees without thinking—has to have had affected employment, but accounting for its subtle, long-term effects is difficult if not impossible. The effect of robots on employment will be difficult to measure, particularly because it cannot be calculated in the absence of robots' human pilots, mechanics, programmers, or other tenders. Raw numbers will be of limited value and will need to be compared to alternative estimates, especially at the level of GDP and national employment figures, where there can be no control population.

But it's important to underline the key finding to date: unlike farmers who moved to the cities as agriculture industrialized, we have few visible indicators of the effect of computers, much less robots, on work and employment. Do robots increase unemployment? Nobody really knows. Erik Brynjolfsson and Andrew McAfee of MIT, in *Race Against the Machine*, suggest that the sheer speed and wide influence of digital innovation have prevented most people (skills and knowledge grow slowly) and organizations (business tasks and processes haven't changed fast enough either) from keeping up with the pace of change.[2] Thus the "jobless recovery" from the 2008 recession cannot be blamed on information technologies—but neither can they be said to have played no role.

According to one school of thought, information technologies coincide with—and may be responsible for—a "hollowing out" of the U.S. workforce. Stagnant middle-class wage growth is well documented and likely has many contributing causes. One of these can be observed as computers take on more and more complex tasks, displacing people who formerly performed them within a job description. Widespread outsourcing of payroll, an essential function that delivers no competitive advantage even when done well, to ADP and other firms has caused the nearly complete demise of the payroll clerk, for example.

David Autor, an economist at MIT, contends that when a job task is new, people are needed to perform it because humans can adapt, analyze, and improvise. As the task gets better understood and codified, machines can take over. The "hole in the middle" refers to vulnerable jobs that are neither low wage and highly physical nor high wage and highly cognitive.[3] Unfortunately, workers whose jobs are displaced struggle to find alternate employment in some new category. Experience shows that such workers are neither geographically portable (sometimes because of family ties, or the effect of underwater mortgages) nor able to land jobs for which their skills might be useful but their professional vocabulary, personal network, or earnings expectations are not. Downward mobility requires difficult adjustments.[4]

Will New Robots Displace More Jobs Than They Create?

The economic theory traditionally invoked in these situations suggests that labor-saving innovations free workers to perform work that adds more value: farmers could stop tilling the soil by hand on small plots when first horses and later tractors made it possible for them to farm ever larger fields. Today, roughly 2 percent of the U.S. population not only feeds the other 98 percent but exports a significant quantity of crops and foods as well. Such a ratio would be inconceivable 100 years ago.

MIT's Autor makes an important point: just because a task *can* be automated does not mean it *will*. Within the same industry, and indeed the same company, automation depends on labor economics: Nissan Motor Company uses more robots at its factories in Japan than it does at those in India, where labor is considerably cheaper.[5] As of 2013, Japan's unemployment rate was 4.0 percent, versus 7.4 percent in the United States. Meanwhile, as of that same year, the International Federation of Robotics reported that Japan had 323 robots per 10,000 workers, whereas the United States deployed 152 per 10,000 workers in the general workforce—up from 72 only ten years prior.[6]

Thus, at the macro level, Japan had more than twice as many robots, proportional to the workforce, but roughly half the unemployment as compared to the United States. Judging from this example at least, it would seem to be hard to prove that robots necessarily give rise to higher unemployment.

But the Japanese economy does not resemble the U.S. economy in most respects. Ethnic diversity, population density, extractive industries (mining, farming, fishing, and energy), and ratios of imports to exports differ considerably between the two countries. Japan is aging more rapidly and admits far fewer immigrants. And attitudes toward robots in the two countries are conditioned by extremely different cultural milieus. Thus it would be premature to cite Japan as definitive proof that robots do not contribute to higher unemployment. It is also easy to envision a scenario in which robots constitute a "reserve army of labor," maintaining downward pressure on wages: those Indian auto workers have a strong disincentive to strike knowing that Nissan has robots in waiting for when wages get sufficiently high.

Factories

How have robots been used in the U.S. workforce thus far? Given that the auto industry is the documented leader in the utilization of robot workers, looking at that sector can be suggestive for larger patterns—or the absence thereof.

To date, robots behave very much like machine tools, doing repetitive tasks for which they are programmed: moving heavy items, spraying paint, or installing components on an assembly line. Robots are usually bolted down, heavy, special purpose, and caged for human safety.

Several emerging trends suggest adoption of a new generation of robot workers. Rethink Robotics' Baxter robot came to market in 2012. Unlike traditional industrial robots, it is relatively cheap (about $25,000), safe around people, easily programmed, and multifunctional. Its target market is small businesses, where the robot can free workers to do more interesting and more valuable work, as opposed to picking items off an assembly line and putting them in bins or shipping boxes, for example. In addition to freeing people from repetitive, boring jobs, Baxter is designed to be installed within human workflows: unlike assembly-line robots, which are capable of inflicting potentially lethal force, it senses contact and can avoid harming humans.

Supply Chain

Another trend can be seen in supply-chain robotic systems, the most visible example of which is made by Kiva, a Boston firm acquired by Amazon in 2012 and now called Amazon Robotics. Kiva is a perfect example of human-robot partnership: humans' eyes and brains can detect patterns far better than robots, and their hands combine touch, adaptability, and nimbleness in ways that far exceed the capabilities of current robot graspers. Robots, on the other hand, are better at repetitive tasks, moving large items, and following bar code stickers on the floor in a prearranged pattern. Thus Kiva never touches an item but, instead, moves racks of retail goods from filling to storage areas and from storage to picking and packing areas. Human workers don't need to walk the long distances characteristic of large distribution centers; the robots don't have to do hard tasks like visual discrimination or small-item picking.[7]

Given how unpredictably Amazon behaves, it is significant that the company has not always deployed Kiva systems in the rapid expansion of its global distribution center network. One possible explanation for the apparent incongruity (other than that large corporate acquisitions take years to sort themselves out) is that Amazon was more interested in the software sophistication behind Kiva than in the mechanical devices themselves.[8] An article from 2007 made this point: Kiva warehouses are self-tuning, in that slower-selling items are moved into less accessible locations, whereas higher-velocity items are stocked on the edges of the storage area. Because the Kiva pods work twenty-four hours a day, moving slow-selling stock during periods of downtime, for example, can be driven by software rather than a human manager trying to juggle multiple priorities. The article's title—"Random-Access Warehouses"—made precisely this point.[9]

The Big Picture

One reading of economic history suggests that when a task gets mechanized or automated, workers find new ways to be involved in the workforce. To take one notable example, as of 1970, one-third of women in the United States workforce were secretaries. With the introduction of the personal computer and word processing software, the need for secretaries fell off dramatically, but the overall number of women employed increased.

In 1992, Robert Reich (who later became secretary of labor in the Clinton administration) predicted the emergence of a three-tier labor market in developed economies.[10] Reich started with personal service jobs (in health care and retail, to take two huge examples), to which he added a second tier, production workers in a diminishing manufacturing sector, and foresaw the rise of a third, what he called "symbolic analysts." This last tier includes financial services, engineering, software, and law. After a surge in its numbers, it is being automated: big data tools are replacing humans, who are not as good as machines in scoring credit ratings or interpreting mammograms. As the *Economist* put it in May 2013: "Bank clerks and travel agents have already been consigned to the dustbin by the thousand; teachers, researchers and writers are next."[11] Accounting and law are getting both offshored and automated: the task of legal discovery used to be labor intensive and thus both expensive (for the client) and profitable (for the partners who put associates and paralegals to work billing the mountain of hours). Now much of the work can be done by software.[12]

Since Reich made his predictions, the degree of U.S. income inequality has increased (see figure 7.1). Gains by the top 20 percent of the population are largely driven by gains to the top 5 percent, or even 1 percent: a household of two public school teachers can make $140,000 between them, but that middle-income group is not the driving force here. Instead, a number of economists argue that large income gains, and thus large increases in income inequality, result from larger returns to capital rather than labor.[14] The shift toward growth in investment-related income that leaves wage-related growth behind coincides closely with a growing gap between wages and productivity since about 1970. In other words, investments in capital such as computers and other forms of automation drove increases in productivity, whose benefits accrued largely to the owners of the capital rather than to workers (see figure 7.2).

Put those two trends together—automation of increasingly complex tasks and increasing returns to capital

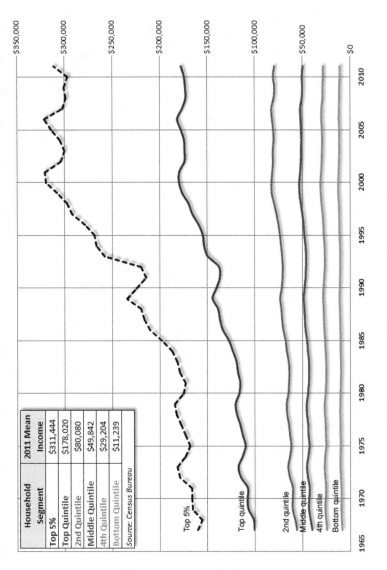

Household Segment	2011 Mean Income
Top 5%	$311,444
Top Quintile	$178,020
2nd Quintile	$80,080
Middle Quintile	$49,842
4th Quintile	$29,204
Bottom Quintile	$11,239

Source: Census Bureau

Top 5%

Top quintile

2nd quintile

Middle quintile

4th quintile

Bottom quintile

Figure 7.1 Inflation-adjusted U.S. mean household income by quintile and top 5 percent (1957–2011).
Source: Census Bureau.[13]

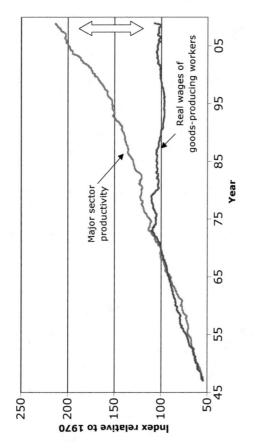

Figure 7.2 Wages stopped tracking productivity growth in the United States in the 1970s. *Source:* Bureau of Labor Statistics.[15]

relative to wages—and the robot would appear to portend bad news for workers. Four developments underlie this negative scenario. First, thanks to economies of scale, learning curves, and Moore's law of microprocessor performance, robots are becoming cheaper every year. Second, given advances in software engineering, machine vision, and other components, along with trickle-down innovation from the high levels of defense-related research and development, robots are also becoming more capable every year. Third, even though wage increases have been minimal, the continued growing rise in health insurance and other expenses incurred by human workers (including the cost of air-conditioning, for example) mean that humans are becoming more expensive every year.

And, fourth, as Illah Nourbakhsh of Carnegie Mellon points out, robots don't need to *replicate* human performance; they just need to be *good enough*. Working alongside robots, humans will learn how to make best use of robots' strengths and how to best cover for their weaknesses (by providing nonmachine vision when a robot gets stuck, for example, or by designing self-checkout lanes where customers do most of the work, with a single employee supervising six registers instead of operating one.) As we climb the learning curve, business processes will be redesigned around the humans' and robots' respective strengths.[16]

Given that low-wage laborers and the unemployed often lack both human and financial capital, it's most likely that robots will first be owned by capital rather than by labor. Higher up the wage scale, it's not as straightforward. Quantitative financial services investors can multiply and amplify their expertise by encoding it in computers, trading networks, and other robotic technologies. It's quite possible that radiologists could be early adopters of mammogram-screening software, keeping control of the tools within the guild, as it were.

Other forecasts are more sanguine: there has always been enough to do to keep most of the workforce busy. Robots will be used for the jobs people didn't want to do anyway: the "three Ds" of dull, dirty, and dangerous tasks are frequently cited. Experience bears this out: there have been successful tests or deployments of robots in bomb disposal, rescue efforts during a Fukushima-like disaster,[17] repetitive assembly-line tasks, even in vacuuming the living room. One school of thought foresees humans having much more leisure time in which to explore and express their interests and talents. Kevin Kelly, a founding editor at *Wired*, wrote in this vein,

> We need to let robots take over. They will do jobs we have been doing, and do them much better than we can. They will do jobs we can't do at all. They will do jobs we never imagined even needed to be done. And they will help us discover new jobs for ourselves, new

tasks that expand who we are. They will let us focus on becoming more human than we were.[18]

Academic roboticists, when they express an opinion, often fall into this camp. Henrik Christensen of Georgia Tech is well regarded in his field, though no labor economist. He asserts, without evidence, that every manufacturing job brought back to the United States from offshore (a move made possible by robotics in many cases) also generates "1.3 other jobs in associated areas."[19] The advantages of robots—lower costs, higher precision, and support for the requirements of "clean rooms" and similarly difficult environments for humans—make the number plausible: even when products are made by robots, the manufacture of these products still needs to be supported by human processes such as procurement, accounting, repair, and marketing.

But the jobs question is not merely a matter of quantity. Frank Levy and Richard Murnane are labor economists who study the effect of information technologies on work. They suggest that, to address "the new division of labor between people and computers," there are "four fundamental questions" we need to answer:

1. What kinds of tasks do humans perform better than computers?

2. What kinds of tasks do computers perform better than humans?

3. In an increasingly computerized world, what well-paid work is left for people to do both now and in the future?

4. How can people learn the skills to do this work?[20]

Displacing an assembly-line worker who put handles on the car door does not free up a nurse to help address a shortage in that field, or a robot programmer, maintainer, or even cleaner. That is, the skills being replaced may well relate to jobs that did little to enhance human dignity in the first place, but at least they were jobs and the path from displacement to redeployment is not at all well defined. Employers frequently invoke a "skills gap" and put pressure on schools and universities to update curricula; it's a valid concern. But Peter Cappelli from the Wharton School of the University of Pennsylvania also directs attention to employers that are reticent to train workers.[21] (There's also the question of how much employers' heavy reliance on deficient résumé-screening software may be driving artificially high unemployment.)[22] Thus robotics' effect on employment has many second-order effects, and it's going to be difficult to get clear answers to what these might be any time soon.

Because there are few one-for-one equivalencies in any area of economics, particularly in the area of employment,

Robotics' effect on employment has many second-order effects, and it's going to be difficult to get clear answers to what these might be any time soon.

we can defer the issue of robots' job displacement for two reasons. First, the confusion over what constitutes a robot makes matters difficult; almost anything that relates to a tool or artifact of any sort might qualify. Second, it may be that job displacement by robots is farther along than most analyses would suggest. As of mid-2013, there were 11.76 million "unemployed persons" versus 143.9 million people who were employed. Simple math puts the mid-2013 unemployment rate at about 8 percent; technical adjustments produced an official rate of 7.6 percent. What is not counted in these numbers are people who gave up looking, people who were working part-time when they wanted and needed to work full-time, and retirees, some of whom left work involuntarily. Another group of people—14 million of them each month—who are not working collect disability payments. The rate of disability filings has nearly doubled since 1996; more than one-third of those claims relate to back pain and musculoskeletal problems and another one-fifth relate to mental illness and developmental disabilities—all conditions difficult to diagnose with any degree of confidence.[23]

Labor economist David Autor argues that disability is "a kind of ugly secret of the American labor market. Part of the reason our unemployment rates have been low, until recently, is that a lot of people who would have trouble finding jobs are on a different program."[24] The fact that the rate of disability filings doubled in a period that coincided with heavy technological unemployment—offshoring, call-center automation, self-service retail—suggests that as Brynjolfsson and McAfee have argued, new technologies are creating wealth and productivity, but not enough jobs for those displaced. Adding 7 million disability claims (half of the current total) into the ranks of the jobless would drive the official unemployment rate to a politically dangerous 12 percent, again not counting underemployment or premature retirements.

Just to show how complicated the picture is, some of the people disabled (on paper anyway) by back and related issues could be partnered with robots that could lift and move things. The scenario raises a thorny question: What will individuals with lifting limitations and often only a high school education bring to the human-robot partnership? Such questions will become more pressing sooner rather than later. In the interim, we are in the middle of a vast experiment involving millions of workers' lives and livelihoods.

HOW DO HUMANS AND ROBOTS GET ALONG?

To date, most robotic research and engineering efforts have been oriented from the inside of the robot out, addressing difficult problems of wayfinding, actuation, grasping, machine vision, and so on. Now that robots are beginning to enter human territory to perform their tasks, a new set of issues arises. What are the rules of the road for humans to steer clear of mobile robots, press elevator buttons for them, or warn them of impending danger? Which party does what part of a task, whether simple (ATM transaction) or complex (bombing runs or surgery)? Where do blame, liability, and ultimate responsibility lie—possibly in the form of a kill switch? A brief sampling of actual scenarios illustrates the potential richness of human-robot partnership as well as the complex issues still to be resolved.

Human-Robot Interaction

Research into human-robot interaction (HRI), which receives substantially less attention than do other technical challenges of robotics, is focused much more heavily on how robots "read" human input than on how humans respond to the presence of robots in workplaces, during emergency rescues, or when public safety is threatened. For example, in a 2013 literature review, Robin Murphy and Debra Schreckenghost, two of the field's most respected researchers, noted that

> in practice, metrics for [human-robot] system interactions are often inferred through observations of the robot or the human, introducing noise and error in analysis. The metrics do not completely capture the impact of autonomy on HRI as they typically focus on the agents, not the capabilities. As a result the current metrics are not helpful for determining what autonomous capabilities and interactions are appropriate for what tasks.[1]

Which is to say, HRI researchers have yet to propose, much less agree on, standard measures of how humans and computers interact. Out of forty-two metrics identified in the literature review, seven apply to humans, six to robots, and twenty-nine to the interaction between them. Of the seven human metrics, only one—trust—might be said to measure the response of a human being approached by an autonomous robot; the others, such as "productive time versus overhead time," apply to a human operating or in control of a robot.[2] In the *Springer Handbook of Robotics*, an encyclopedic collection of articles by the field's leading scholars, the authors of "Social Robots That Interact with Humans" acknowledged that study of robots' effects on humans is in an early stage. They posed a question, one central to the field of HRI research, that remains to be answered: "What are the common social mechanisms of communication and understanding that can produce efficient, enjoyable, natural, and meaningful interactions between humans and robots?"[3]

The Case of Search and Rescue

Robots are ideally suited to the often dangerous and dirty search and rescue process.[4] Although there are some surprising considerations in the design of search and rescue robots, the humanitarian upside of this type of robot stands in clear contrast to other types, where the moral and ethical dimensions are more complicated: industrial robots take away some people's livelihoods, combat robots are already raising serious issues (quite literally of life and death), and even care robots run the risk of dehumanizing nursing home residents. Indeed, it's hard to find a downside to robots that save human lives in dangerous circumstances. In such a straightforward mission, however, there are numerous questions in the handoff between human and robot that must be addressed.

The field of search and rescue is as broad as human activity itself: search and rescue robots are currently being tested in the air, on the ground, and both on and under water. A landslide or tsunami requires aerial robots to assess broad geographic expanses; buildings badly damaged or demolished by fire, earthquake, or explosion require ground robots to crawl through rubble. Even rubble itself is so heterogeneous that, as of 2006, there was no technical standard to characterize the different detritus left by a fire, earthquake, and explosion.

Search and rescue robots must also be specialized by activity: sensing and assessing a damaged building's structural integrity; sniffing and identifying gases (explosive or poisonous); detecting and measuring radiation and mapping contaminated areas; locating, assessing, delivering care to, and extricating survivors; and mapping various layers and scales of terrain affected by a disaster each require different designs, operators, and protocols. There's also the question of getting the robot to the disaster area in the first place: hauling in a 1,000-pound machine from hundreds of miles away may present a real problem when transportation to and communication with the area are at a premium.

Thus far, search and rescue robots have most successfully been used in the air, for mapping and other reconnaissance tasks. Airborne runs are easily scheduled (especially at altitudes below those used by manned aircraft), aircraft with internal combustion engines can usually be deployed, and airspace presents fewer unexpected obstacles. In rubble, by contrast, battery life is an issue, especially when the robot encounters unexpected obstacles. The lack of wireless bandwidth underground or in concrete-, stone-, or steel-intensive rubble presents a major difficulty, often necessitating use of a fiber-optic tether and safety rope (both of which get readily snagged in disaster debris). Particular conditions will slow down a given robot's locomotion: protruding sticks or rebar can detread open-side tank tracks; even shag carpet can present a serious problem: indeed, it proved to be the undoing of one search and rescue robot after a mudslide. Ash and water from fire hoses can combine to make surfaces extremely slippery and also to obscure the robot's camera lenses. By reviewing some of the design challenges that confront search and rescue roboticists, we can gain an appreciation for the vast potential of this field.

Rules and Algorithms

Firefighters, police officers, and search and rescue teams all follow guidelines for how to approach a chaotic, dangerous situation. Which rooms are searched when, what safety measures are called for by what hazards, and what communications actions are required—all of these are inculcated by training and experience. Teaching a robot how to behave in a scenario that will be unique, confusing, and dangerous is a significant challenge. Balancing autonomous and user-directed behavior, for instance, is critical. When UAVs were used to assess the structural integrity of midsize commercial buildings after Hurricane Katrina, for example, ground-based operators flew the drones, but having an autonomous wall-standoff capability would have been desirable to reduce stress on the operators, who were in line of sight with the UAVs but still faced challenging wind conditions close to the damaged structures. Well-intentioned robot designs have repeatedly been rejected in disaster scenarios when responders failed to intuitively grasp how to operate the devices.

Setup and Maintenance
How fast can the search and rescue robot be unpacked? How long does it take for an inexperienced operator to change batteries? Where is the latest operating manual? Internet downloads are often a great solution, but not when there's neither cellular service nor electricity, much less a printer. What language(s) are the instructions written in? Designing reliability into a robot that will encounter dust one week, mud another, extreme cold a month later, then sit in a hot warehouse for a year is a significant

challenge: in very few projects will so many operating parameters be unpredictable.

Where Am I?

Disasters rearrange the landscape at multiple levels. An urgent task early in disaster response is to integrate what is previously known with what is discovered. Has this building been searched? Is this bridge safe to walk, motorcycle, or convoy over? Where are the gas lines? Are they shut off? Developing sensors, data standards, and related rules of engagement for spatial information as both used and collected by robots is another high-priority objective not easy to achieve.

Mobility

Wheels, legs, treads, wings, and propellers each have strengths and weaknesses. Knowing only that the environment will be harsh and physically challenging makes deciding on the best mode of locomotion for a robot very difficult. Invertible robots are desirable in tight spaces where backing up to turn around is often impossible and where human operators' sense of the interior landscape makes accurate remote control more difficult. Legged and other biomimetic robots are capable of greater mobility in unpredictable settings but present difficult engineering challenges; snakelike robots advance well in highly uneven rubble but are still difficult to build.

One promising area of search and rescue robotics lies in the use of teams, either of robots and humans (with or without dogs) or of robots and robots. A helicopter or blimp overhead can survey wide areas so ground robots can be informed how close they are to a live electrical line, a pier or other drop-off into water, or to territory already searched. The eyes in the sky complement the sensors in the rubble. And though no robot can approach a dog's olfactory acuity, sniffing robots can be deployed in hazardous situations where human handlers are unsure of the risk to their canine partners. Swarms of robots that set up ad hoc mesh wireless communication can offer redundancy in the event one member of the swarm is incapacitated and can cover large areas in parallel rather than in series. Discovering how many human operators are required to manage groups of robots, however, remains an ongoing challenge.

Structure

Given that much of the world's population lives near water, that water has destructive power in several forms, and that water is almost always involved in firefighting efforts, how important is it to waterproof a ground robot? There is usually a structural trade-off between a robot's ease of access for maintenance and its resistance to damage from water, sharp objects, or other hazards. How many rescue personnel are required to carry, launch, and retrieve the robot? Weight, battery life, and capability constitute difficult trade-offs. Current search and rescue robots cannot move anything very heavy; although military rescue robots can drag a wounded soldier to safety, provided the soldier's wounds are not severe, neither military nor civilian rescue robots are as yet capable of safely transporting humans who need spinal stabilization, a common requirement in building rubble.

Roles and Modes

A more complex issue of physical structure relates to how the search and rescue robot interacts with its operator(s), with dogs and rescue personnel working alongside it, and, even more important, *with the person being rescued*. What used to be called "user interface" grows far more involved when a robot interacts with multiple people in multiple roles; human-robot interaction is of particular importance in civilian search and rescue robots. A military rescue robot would likely have a dedicated operator and support team who would have trained with one another, approaching soldiers who know how to be rescued. In contrast, a rescue robot and its civilian support team on the scene of a disaster may not have trained with local responders, and would find themselves in search of a civilian who most likely had *no* mental framework or training in how to be found or extricated by a robot.

What information do operators need? Visual feeds from rescue robots' camera(s) can be useful, of course, but the operator cannot be assumed to be sufficiently safe and undistracted to focus only on the monitor(s). One school of thought suggests operators should have the robot's-eye view of its physical context, information relating to the robot's operational condition (battery life, operating temperature), and a bird's-eye view of the robot's location in the disaster area. To properly manage this much information, search and rescue robots probably need multiple operators, for a variety of reasons involving the performance, endurance, and emotional state of human operators. One study suggests that a second operator improves the rescue robot's operational performance by a factor of 9.[5] Add to this the support team, and the ratio of humans to robots becomes an important consideration.

Toward a Continuum Model of Interaction

In contrast to binary debates over "whether X is or is not a robot," new efforts focus on a continuum of human-robot partnerships in the gray area between two extremes, one actual and the other theoretical. Take as an archetype of one extreme a newborn human, a purely biological creature, without language or much in the way of cognition. At the other extreme, consider a purely artificial, disembodied creation, an archetype that in fact may never be realized, like HAL 9000 in *2001: A Space Odyssey*, which possesses sense and logic and which can act (turn off life-support systems, lock outer doors) but not move.

The interesting ground for defining human-robot partnerships is the vast conceptual area that lies between these two extremes, combinations of sensors and sense organs, cognition and logic, and action by either bones and muscles or hydraulics and motors. Key questions arise about assistance, capability, and responsibility within these partnerships. Which partner assists the other, and in what ways? Which partner is ultimately in control? And what can a particular partnership achieve that neither partner could alone?

Two brief examples illustrate the usefulness of the hybrid approach to human-robot partnerships.

1. When a customer retrieves money from an ATM, the human provides key capabilities to the robot, making the customer–cash machine interaction a human-robot partnership. The same goes for smartphones or other GPS navigation systems. The computer responds to its human partner's request, senses where the human is in space, calculates a route, and relies on the human to follow directions to fulfill that request.

2. When humans receive biomedical augmentation (cochlear implants, robotic arms, or wheelchair/speech synthesizers like Stephen Hawking's), however much

the human-technology partnership relies on robotics, there is never any question about the human partner's humanity—and agency—within the partnership.

There are many gray areas in human-robot partnerships, whether they derive from carbon-fiber running prostheses, Google Glass facial recognition, or automated stock-trading algorithms. Instead of worrying about binary definitions, informed and nuanced debate can help us clarify the place and limits of both augmented humanity and humanized automation.

To suggest but one question that might be raised in this debate, how soon will there be "unlimited" categories of athletic competition, much as there are in competitive motor sports? Exoskeletons, prostheses, and various implants might be legalized for a new competitive class of augmented human athletes.

Note that, once the discussion of human-robot partnerships puts humans and robots on a continuum, the door is open to pharmaceutical or medical augmentation. Steroids, human growth hormones, and blood transfusions can improve human musculoskeletal performance through a variety of mechanisms. Beta-blockers have long been recommended (or self-prescribed) for people with strong adverse reactions to speaking or performing in public. New medications for post-traumatic stress disorder

(PTSD) might enable people who relive or dream about their traumatic experiences to forget those experiences. Each year, tens of millions of prescriptions are written for attention deficit/hyperactivity disorder (ADHD) medications; many of these pills find their way to nonsymptomatic users, whether to enhance their moods or improve their performance (students cramming for finals, athletes wanting an extra edge). Professional football players have been suspended in the United States for taking one such medication in particular, Adderall.[6] The point here is that the discussion about human augmentation in many competitive scenarios lags far behind the reality of the various legal and illegal means available to achieve it. Robotic augmentation is in some ways merely one category among several.

Compu-Mechanical Augmentation

One category of human-robot partnership entails compu-mechanical assistance to a human, possibly disabled, often not. Beginning with the lever, machines have augmented and then largely replaced human *muscle* in growing and harvesting crops, extracting natural resources, and fabricating the man-made environment. Beginning with the stylus, most likely, a succession of tools have aided and largely replaced human *brainpower*: a high schooler with a $2 calculator is far better at doing simple math than a Ph.D. without one; computer-augmented social networks can help solve crimes, predict elections, and solve complex problems. The robot combines these two varieties of human assistance and augmentation, musculoskeletal and cognitive.

Put a slightly different way, machines multiply force. Computers, singly and in networks, multiply and amplify cognition. Once computing moves into three dimensions, robotic technologies multiply *presence*, allowing humans to sense, observe, analyze, and act on physical reality at some remove. In 2012, Steve Cousins, then CEO of the humanoid-robot start-up Willow Garage, envisioned a near-term future in which his company's video conferencing robots would transcend telepresence, allowing humans to accomplish actual tasks remotely rather than merely to see what's going on.[7] Asking, "What is the nature of the human-robot partnership?" lets us dig more deeply into questions of costs, benefits, and risks; resource allocation; ethics, particularly agency; and other issues that thus far have been left largely unaddressed.

The Case of Surgery
By far the most visible robotic treatment technology, the da Vinci Surgical System is really not a robot: it can neither move nor perform any movement autonomously. Much like the prosthetics described below, however, the

Asking, "What is the nature of the human-robotic partnership?" allows the discussion to dig more deeply into questions of costs, benefits, and risks.

da Vinci's robotic enhancement of human capability is sufficiently powerful that it deserves our attention here. Furthermore, a robotic technology that is prominently featured in hospitals' marketing efforts (including billboards) merits attention if only because the robotic nomenclature is helping to shape public discussion of the field.

The da Vinci was developed in conjunction with a military initiative to test the feasibility of removing surgeons from medical facilities close to the front line while getting wounded soldiers aid as quickly as possible. Although that design goal was abandoned, it led the way to development of robotic arms with sensors, instruments, and various tools under the control of a surgeon seated at a console.

When the da Vinci came onto the market in the late 1990s, it was claimed that robot-assisted surgery marked a third stage of surgical procedures, after open surgery and minimally invasive techniques such as laparoscopies. Though also a form of minimally invasive surgery, unlike some older types, the da Vinci allows the surgeon to operate in actual-image rather than mirror-image mode, so that moving the joystick at the console to the right, say, means the instrument in the operating field goes to the right as well. Thus the robotic augmentation functions as an extension of the surgeon's eyes and hands.

In the market for more than a decade, the da Vinci provides some useful economic data for other robotics companies still exploring alternative business models. The device itself sells for between \$1 million and \$2.3 million, depending on geography and configurations. In addition, the system has components that wear out and accessory products such as tips that must be replaced after each surgery, so that the sale of consumables and spare parts generates ongoing revenue (in economic terms, these sales constitute a form of lock-in given that there are no alternative providers to compete with on price). Finally, a service contract of between \$100,000 and \$170,000 per machine per year also applies. To give a sense of scale, 449,000 da Vinci procedures were performed in 2014, compared to 367,000 in 2012; the installed base of da Vinci Surgical Systems was reported to be 3,266 as of December 31, 2014.[8]

There are considerable economic incentives for Intuitive Surgical to have its devices used as much as possible; this is despite the lack of documentation that better outcomes accompany the higher cost of robot-assisted surgery.[9] Thus prostate surgery is a high-volume da Vinci procedure, but, according to the *Journal of Clinical Oncology* (2012), incontinence and sexual impotence are high after both robot-assisted and conventional laparoscopic surgeries. And the *Journal of the American Medical Association* stated in 2013 that "to date, robotically assisted hysterectomy has not been shown to be more effective than laparoscopy." Partly because of the high public profile of the da Vinci Surgical System, however, hospitals charge up to twice as much for da Vinci as for conventional surgery, and insurance companies, to date, have paid more for the process, again with no documented improvements in outcomes.[10]

Prostheses

Progress in the field of active prosthetics relies on developing all three components that define a robot: sensing, logic, and action. Progress for each of these components has advanced to the point where mind-controlled robot prostheses have been demonstrated. In particular, sensors that detect nerve impulses in the amputee patient's stump can control prosthetic arms, hands, and legs. Research and development in this domain are under way in many countries, including Israel, Sweden, the United Kingdom, and the United States.

The wars in the Middle East in the first decades of the twenty-first century were marked by impressive advances in the care of wounded soldiers: compared to soldiers injured in Vietnam, where only 76 percent survived, soldiers injured in Iraq had a 90 percent chance of survival, even though there were fewer corpsmen and doctors per patient.[11] That high survival rate, however, came at the price of thousands of amputees, victims of the improvised explosive devices, mines, and other tools of asymmetric warfare. Given that these young amputees would otherwise face the mental and physical hardships of crutches or confinement to a wheelchair for perhaps sixty or more years, developing effective prostheses that utilize robotic technologies—including brain-signal interfaces[12]—is a high priority of current research.

In addition to robotic appendages, exoskeletons like ReWalk for patients with intact limbs but limited function (specifically, paraplegics) can help them walk. In 2012, a paralyzed woman walked the entire London Marathon course in 16 days using a ReWalk. The system, which currently costs approximately $85,000, is approved for use in medical facilities under the supervision of a physical therapist or similarly skilled individual. In the future, patients may be able to use the devices at home. For less seriously impaired individuals who have weakened muscle function, Honda has developed Stride Assist and Bodyweight Support Assist for help in walking, but it's not clear when the devices will become commercially available.

Assistance with Daily Life

Robots that assist people with the tasks of daily living help promote independence, particularly for elders without family nearby. Although the capabilities of assistive robots are important, so are the attitudes of those receiving assistance. A recent study at Georgia Tech suggests that physically impaired people in the United States welcome robotic assistance with tasks like cleaning but still prefer humans to help move, feed, and bathe them. Caregivers, for their part, clearly preferred working with robotic assistance to being replaced altogether.[13]

One robot designed for home assistance is the Yurina from Japan Logic Machine. Announced in 2010, the robot can lift light adults out of bed, carry them (as to a tub), or serve as a motorized wheelchair. As with other medical robots, however, when the Yurina will become commercially available is unclear.[14] Given the hazards of lifting patients, such assistance would seem to have a potentially wide appeal among caregivers.

The Bestic feeding assistance robot is produced by a Swedish firm whose founder was limited in his arm function after contracting polio as a teenager. Because eating is so important to social interaction, being able to be self-sufficient at a meal contributes to well-being at several levels. The Bestic sits on the tabletop and has a clean, white design. It can be controlled by foot pedals, buttons, joystick, and potentially voice. The My Spoon, from Japan, performs similar functions.[15] Personal feeding robots are highly culture specific, given the wide differences in table utensils, food textures, and eating customs around the world.

One factor in the choice of technological assistance lies in the embedded psychological signals that accompany a given technology. Sitting in a wheelchair, no matter how robotic, means not being able to look standing adults in the eye. Robotic walking devices such as exoskeletons change patients' attitudes (in both senses of the word) toward their surroundings.[16] Many older people need a bit of help getting out of a chair, and the French company Robosoft's robuLAB-10 has been shown to perform this task effectively. Along with several other robotic devices, the robuLAB-10 is intended for use in institutions such as rehabilitation hospitals, but the market is developing slowly. It is not difficult to envision such devices in a residential setting as software improves, safety measures are tested in wider trials, product liability is clarified, and other barriers to wider adoption overcome.

Given the rate of increase in the aged population in so many industrialized countries, and given developments that can cross-pollinate multiple areas of robotics (better motors, shared software libraries, new materials, mind-computer interfaces), the pace of innovation in robots that improve self-sufficiency should be rapid.

Monitoring

Robots used in the care of the elderly often perform multiple functions. Although the GeckoSystems CareBot does not deliver care, it can observe individuals and provide feedback about their behavior, or remind them to eat, take a medicine, or let in the cat. From a Swedish university comes Giraffe, an elder-assist robot that can monitor blood pressure, note a person's movements (and learn the person's regular sleep patterns), and send an alarm if the person should fall or become immobilized.

Companionship

A quick look at age pyramids for industrialized countries indicates a fast-approaching predicament: how to care for a population of elders who, because of improved diet and health care, live longer than ever when there are proportionally fewer working-age people to support a growing non-wage-earning population? Japan presents a pressing case: the proportion of persons over the age of 65 grew from 5 percent of the population in 1950 to 23 percent in 2010, and could reach 40 percent of the population in 2050.

On the one hand, economic productivity among the working-age population in Japan and many other industrialized countries must improve in order to fund the growing number of retirees; savings alone will be insufficient. On the other hand, to supply nurse's aides and other caregivers in current proportions would create labor shortages and economic imbalance. Enter both the carebot, to help relieve the demand for caregivers, and more industrial and service robots, to reset the economies of countries with rapidly aging populations.

The Paro robot is a stuffed animal modeled on a baby harp seal (see figure 8.1). It was developed by the National

Figure 8.1 Paro elder-care robot. Photo: University of California, Irvine, Creative Commons license.

Institute of Advanced Science and Technology (AIST), a Japanese public research organization, at an estimated cost of $15 million, according to the *Wall Street Journal*.[17] After being launched in 2003, Paro robots are in their eighth generation and cost approximately $6,000 apiece. The Paro was built on the premise that the benefits of animal therapy could be delivered without the complications of maintaining large numbers of animals in an institutional setting.

The Paro is equipped with five types of sensors:

- tactile

- light

- sound

- temperature

- posture

A robot in the strictest sense of the term, the Paro can distinguish light from dark and thus also sleep cycles, both its own and a human's. When the human strokes or speaks to it, the Paro can detect intention and respond accordingly with gestures and sounds. The robot's body and facial expressions are evocative and some elders, particularly those with certain forms of dementia, are reported to be calmed by the device.

For something designed to be so cute and cuddly, the Paro has proven to be controversial. Some critics point to the inauthentic nature of having people "love" an inanimate object.[18] Others express concern that, if the elder embraces the Paro, caregivers and especially the elder's children will no longer feel the need to provide meaningful human contact, telling themselves instead, as one article on the topic put it, "'Don't worry about Granny, she's got the robot to talk to.'"[19]

Centaurs

In some cases of human-robot partnership, the human and the robot augment each other, but a division of labor between the partners that is close to fifty-fifty, though promising in its possibilities, turns out to be both complex and hard to achieve. The vision of such partnership is well formed in the academic literature. University of Southern California roboticist George Bekey wrote in 2005: "We expect a human-robot symbiosis in which it will be natural to see cooperation between robots and humans on both simple and complex tasks."[20] More recently, Erik Brynjolfsson and Andrew McAfee of MIT predicted that "the second machine age will be characterized by countless instances of machine intelligence and billions of interconnected brains working together to better understand and improve our world."[21]

To understand this class of partnership, it helps to ask, "Who performs better: a computer or a human?" The short answer is obvious: it depends on the task. Computers are now unquestionably better at chess than even a grandmaster-level human player, and the highly visible triumph of IBM's Watson over the best *Jeopardy!* players shows how artificial intelligence can be successfully applied to a linguistically rich trivia contest.

What might come next? In early 2015, four of the top ten poker players in the world played a marathon against a Carnegie Mellon computer. Considering the complexities of no-limit Texas hold 'em poker, the researchers were hardly surprised that the outcome was no *Jeopardy!*-like rout, but the statistical tie elated them. Each player played 20,000 hands; a total of $170 million in chips was bet over the two-week competition. In the end, the humans came out less than $1 million ahead—even though the computer did things like betting $19,000 to win a $700 pot.[22]

The "long" answer to the "Who's better?" question is emerging: a team of both. The term "centaurs" usefully denotes a human-robot team in which both team members do what they do best. We are seeing that teams of humans *and* robots outperform *either* humans *or* robots. Here are four areas in which progress is being made more rapidly than might be widely known.

1.   Audi has teamed with Stanford's autonomous vehicle lab to develop a race car that can beat a club-level human driver on time. Given that there have yet to be head-to-head races, neither the adrenaline nor the racing tactics of a competing with purely human drivers have come into play. The Audi simply follows a preprogrammed line and parameters around the course: it hasn't actually raced anyone and won yet.[23] The centaur model is well developed here: stability control, antilock brakes, and sophisticated all-wheel-drive control systems all digitally amplify the

skill of a human driver. Outside vintage models, it's difficult to find cars that are *not* centaur-like.

2. The Internet is awash in images, some of them incredibly beautiful. Researchers at Yahoo Labs and the University of Barcelona have taught an algorithm to trawl through image databases and find beautiful but underappreciated images by using the results of training sessions with human "votes."[24] As the *Economist* noted, the field of machine learning is itself undergoing rapid improvement, in part through the process of "deep learning" as developed by the giant web businesses with both massive data and effectively unlimited computing resources. Google and Facebook are familiar names on their list; the Chinese web services company Baidu is a newer entrant into the field of AI human-robot teamwork, having made some high-profile hires.[25]

3. Chess has never been the same since Deep Blue defeated Garry Kasparov, in part because of a software bug that led Kasparov to infer that the machine was substantially smarter than he, rather than that the computer had somehow made a dumb move.[26] Since about 2013, however, centaur teams of average players and good software have been able to defeat both grandmaster humans and computers. This type of match is where the "centaur" terminology first took hold.[27]

4.   Exoskeletons are common in Hollywood sci-fi, but robots that encase a human body and amplify its capabilities are coming into use in several areas:

- Rehabilitation for stroke patients, amputees, and paralytics.

- Physical augmentation of soldiers so they can march or run longer with less fatigue (DARPA), and of able-bodied humans to increase their lifting capacity, for example (military and other contexts).

- Robotically assisted surgery. The da Vinci Surgical System is a specialized exoskeleton of a sort, extending a doctor's finger manipulations into more precise movements in the surgical field.

One big challenge faced by designers of wearable exoskeletons is to make the power source light enough to work at human scale. In warehouses, a forklift truck typically weighs 1.6 to 2 times the intended weight to be carried. For a 150-pound human intending to carry 200 additional pounds, that ratio would put the human's exoskeleton in the 650-pound range, unloaded, so that the fully loaded package would weigh about 1,000 pounds (a half ton or roughly 450 kilograms). Lowering the battery weight is the quickest way to shrink the weight of the total assembly: a great deal of battery power would be expended in simply carrying the battery and a frame sufficiently robust to support the battery.

It will bear watching to see how roboticists and computer scientists design the cyber side of the centaur, optimizing around human strengths that might be expressed in unpredictable ways. Similarly, training a human to leave part of the task to a machine, and not to overthink the centaur relationship, might be tricky in certain situations. In others (traction control on current cars, for example), humans are already augmented and don't even realize it. When they are explicitly asked in experimental settings, however, humans hesitate to trust a machine's judgment.[28]

At the same time, centaurs will have to deal with both the infinite supply of human stupidity and the limits of algorithmic cleverness. What will self-driving cars do when they encounter a drunk driver headed the wrong way on a divided highway? And what will Wall Street do when programmatic trading robots react in unstable, unpredictable ways to the gambit of a clever day trader? The 2010 "Flash Crash" appears to have been initiated by one person in England who apparently spoofed enough orders—manually rather than algorithmically—to trigger erratic behavior by black box systems that disturbed the entire market. (The gambit seems to have worked, by the way: the day trader made $40 million over four years.)[29] The point here is that the unexpected interactions between stupid or clever humans and fallible computerized entities will be a most complicated territory for decades to come.

Complications

Tracing its roots to ancient efforts to create human capability from inanimate materials, twenty-first-century robotics must be understood in the context of everything from Frankenstein's creature to machine tools. Given this rich and complex legacy, it remains impossible to arrive at a definitive understanding of how robots and humans can and could work together. Pioneering research in a variety of disciplines suggests some promising directions, however.

Unlike other categories of tools, computers that move among people raise significantly different issues than purely mechanical devices do. Two phenomena are of interest here.

Uncanniness

The "uncanny valley" refers to computer animations and robots that are extremely lifelike, but jarring to humans. A classic example of the first, in stark contrast to a low-resolution hand-drawn Disney animation, not very lifelike yet timelessly compelling, is the digital animation of the

Tom Hanks conductor character in *The Polar Express*. Regardless of the number of pixels and the body-motion capture devices used to digitize Hanks's voice performance, certain eye-muscle and other facial motions and shadings of the animation were disconcerting to viewers: gains in technical capability did not translate into greater appeal of the character, even though they had generally done so in previous computer animations.

The same goes for robots. Skin polymers or facial movements that are too lifelike can be off-putting to humans, for reasons that, though evident, are not fully understood. In light of this, Jibo, a "family robot" introduced in 2014, is far less humanlike than its lab-experiment predecessor, Kismet (see figure 8.2, A and B).

The Anthropomorphic Effect

Even before computing left the box in mobile robots, humans were interacting with inanimate objects in surprising ways. The classic work in this regard was done by Byron Reeves and Clifford Nass, who meticulously measured human responses to PCs. Reeves and Nass found that people, across rich and poor, young and old, male and female, routinely assigned human attributes—intelligence, learning, memory, and personality—to robots even as early as the 1980s, and nobody felt compelled to challenge the characterizations, applied as they were to "a collection of wire, silicon, mechanical joints, and computer code."[30]

Figure 8.2 (A) Kismet, an expressive robot developed at MIT. (B) Jibo, a commercial social robot.

The MIT psychologist Sherry Turkle works closely with artificial intelligence, robotics, and other researchers exploring the fuzzy boundaries between humans and machines. She has been eloquently critical of how mobile computing, social networking, and other digital technologies can isolate people and perhaps reshape the human emotional landscape in damaging ways.[31] In short, she is far from a technological drumbeater. Nevertheless, when she was in the lab with Kismet's fellow robot Cog in the 1990s, her own behavior changed:

> Cog "noticed" me after I entered its room. Its head turned to follow me and I was embarrassed to note that this made me happy. I found myself competing with another visitor for its attention. At one point, I felt sure Cog's eyes had "caught" my own. My visit left me shaken—not by anything Cog was able to accomplish but by my own reaction to "him." …
> Despite myself and despite my continuing skepticism about this research project, I had behaved as though in the presence of another being.[32]

Turkle is not alone in attributing human attributes to inanimate objects. In Iraq and Afghanistan, iRobot's battlefield robots saved human lives by helping keep soldiers out of harm's way in the search and neutralization of improvised explosive devices. When the robots were damaged by explosion, the units sometimes had to be shipped back to the iRobot facility outside Boston. According to a 2006 news story, "EOD (Explosive Ordinance Disposal) personnel have made routine use of the units and crafted nicknames and personalities for them. A unit, nicknamed 'Scooby Doo,' earned a check mark on its camera head for each explosive device it succeeded in disarming." When Scooby Doo was destroyed, the article noted, "its operator ... returned it to the repair shop, cradling it in his arms as if it were a wounded child and asking if it could be fixed."[33]

The *Wall Street Journal* reported the same behavior in 2012. Noting that troops sometimes become emotionally attached to battlefield robots, one officer who holds a Ph.D. in robotics noted that "the soldiers and Marines sometimes name their robots—and even give them battlefield 'promotions' for successfully spotting mines or explosive devices." When robots are damaged, "some troops insist that they get the same robot back—not a replacement unit." I have been told much the same story by an iRobot spokesman, and P. W. Singer reports it in *Wired for War*.[34]

A related example comes from the world of entertainment. In the 1980s, the television show *Knight Rider* featured both the young actor David Hasselhoff and a talking

Pontiac Trans Am named "KITT." When the car was later featured at the Universal Studios theme park, lines of people waited to sit in the car and have it speak to them much in the manner of the original mechanical Turk: via a human connected to a remote microphone.

Social Roles

Pioneering research in 2011 at the Carnegie Mellon Robotics Institute placed a snack-delivery robot in an office context, then recorded human reactions to the Snackbot's activities and presence. Participants ordered snacks through a web interface. The wheeled robot, standing about 5 feet tall, had a display on its "head" that expressed emotions and a voice synthesizer that played preprogrammed scripts for greeting, small talk, the snack transaction, and social leave-taking.

Although the human participants were expected to have minimal interaction with a "delivery cart that left snacks," the evolution of their reactions was fascinating. Anthropomorphism was common: people felt sorry for the bot when it broke or talked to a closed door. The robot's nonjudgmental persona was appealing to some, and "he" was accepted as a member of the workplace within two weeks. Norms for interactions with Snackbot emerged; standard human politeness (including not interrupting the bot's speech) replaced machinelike interactions; in one transcript, a coworker told a colleague, "Now you've gone and made Snackbot feel bad." In other settings, participants felt jealous if the bot complimented a coworker's work ethic or healthy snack choice. Snackbot was thought to have "crushes" on several workers on the basis of its speech or travel patterns.

The researchers saw "ripple effects" far beyond the humans' reactions to the bot: the people expressed "politeness, protection of the robot, mimicry, social comparison, and even jealousy." The presence of Snackbot changed how people interacted with one another, in largely unanticipated ways.[35] If a low-functioning snack-delivery machine can have such an effect on humans, how much greater an effect will far more capable robots have on us in the future? And how well will managers, researchers, and others be able to monitor and modulate that effect?

Whether in a laboratory or on a battlefield, at a theme park or in a living room, people involuntarily and consistently react to electronic and mechanical objects in psychologically important ways. But what are they reacting to? There is a core group of both science fiction authors and roboticists who insist that robots can attain consciousness. Rodney Brooks, formerly at MIT, is by no means alone when he writes: "My own beliefs say that we are machines, and from that I conclude that there is no reason, in principle, that it is not possible to build a machine from silicon and steel that has both genuine emotions and consciousness."[36] Much like Ray Kurzweil, Brooks reasons that the continued intermingling of "artificial" and "natural" subsystems will lead to creation of a hybrid life-form and that "the distinction between us and robots is going to disappear."[37] Although that day may never come, the question remains, why do humans react with such intense emotion to robotic forms?

FUTURE DIRECTIONS

However it manifests itself, computing is changing. These changes have important consequences because computing amplifies human cognition—our own and that of those observing and analyzing us. Because we define ourselves much more by what we think and say than by what we do, current robotics is moving close to definitions and assertions of human identity. At the same time, machines powered by those computing capabilities which inhabit the physical world are taking on some traits that can be interpreted or attributed as human. Four broad changes in computing affect people in new ways.

Shape
Whether it is wearables, humanoid robots, self-replicating 3-D printers, or robotic vehicles, what we

think of as a computer has changed forever and made its onetime embodiment in a beige box seem like a distant memory.

Scale

There was much snickering in the 1990s at the quotation allegedly attributed to IBM CEO Thomas Watson that "I think there is only a worldwide market for four or five computers." Now as Google and Amazon build planet-scale networks of data centers, the gist of the statement has a ring of truth to it. Whether we're using Apple Siri, watching a Netflix movie, navigating from a Google map, reading a web-based e-mail, or accessing Facebook, the old PC-centric world of applications, networks, and processors seems less and less in tune with our everyday reality. The idea of a robotic device as a physical manifestation of a global computing grid will grow more familiar in the coming years.

Human Proximity

The "personal" computer sat on a desk, often secured by a lock and cable; smartphones were located closer still, in pockets and purses and on nightstands. But now computers can nest in shoes, eyeglasses, prostheses, and even nerve endings. As humans and silicon-based computing platforms intermingle ever more thoroughly, interesting things will happen, some of them disturbing and others inspiring.

Scope

From manipulating numbers and calculating artillery trajectories, to word processing, to music production, to Photoshop, and now to artificial intelligence, computing has come a long way, moving ever closer to some of the processes that define humanity. Given the pervasiveness and magnitude of these changes, it is important that we examine both what we value and how computing affects our activities and beliefs.

Unlike steam and automotive power, which we came to measure against the power of horses, we have no comparable way to measure how closely artificial intelligence is approaching or amplifying human cognition. A car with 180 horsepower can readily be compared to another with 300 horsepower, but how are we to grasp the relative strength, power, or magnitude of cloud computing, advanced sports measurements, stock-market algorithmic trading systems, or even the relative power of the more tangible voice-recognition smartphone assistants? When Siri 3.0 debuts, how much "better" will Apple say it performs?

This lack of tangibility and yardsticks is important to consider as computing comes ever closer to doing some of the things that make us human. As that computing power is turned loose and comes to both inhabit and measure our physical existence, the need for realistic, informed dialogue grows greater still: computing is now doing humanlike things, in human space, and in conjunction with humans. But we lack language to describe what robots are doing or how well this year's models are doing it relative to 2010, say.

Five particular clusters of issues are arising. In combination, these issues pose significant questions about human identity, agency, and rights and responsibilities. As I said in the introduction, these questions need to be addressed by more people than just computer scientists and engineers.

1. Big Data and Its Insights and Illusions

Turning the physical world into a data model requires massive computing, let alone storage, capacity. One reason robots and self-driving cars are practical is the sensors, algorithms, and processing capacity that can be brought to bear on the task of navigation. In the realm of nonrobotic sensors, meanwhile, surveillance cameras are notorious for producing vast volumes of information (most of it not useful except after human viewing), and the accumulation of machine-generated pips, chirps, and other signals points to a time when the glut of information will overwhelm us—at least until signal processing and interpretation improve. However these fields evolve, robotics will for the foreseeable future be connected with the mythology and technical progress of "big data," defined variously.

2. New Role of Capital versus Labor

As Erik Brynjolfsson and Andrew McAfee argue in *The Second Machine Age*, power laws characterize many aspects of connected systems.[1] In the economy, for example, the richest become richer and more powerful while the least skilled become poorer and more marginalized. Every year there are fewer paths from the bottom rungs of the economic ladder to the top; intergenerational social mobility is slowing in many countries.[2] This growing polarization, and the role that computing plays in it, might explain Google's heavy investments in robotic technologies: having ceded social networking to Facebook (the "next Google"), Google now wants to own the key patents and markets related to physical computing, whether in the car, on the face (Glass), on the wall (Nest), in the factory (JV with Foxconn), or in extreme scenarios (Schaft).

3. Privacy

As robots sense and move among us, they will collect masses of data. As we have seen in multiple data breaches exposing tens of millions of people's information, some of it extremely personal (fingerprints, for example), the scope for the invasion of personal privacy expands every year. To take but one robotic capability, facial recognition is not something we can opt into at Facebook or elsewhere. But with Google Glass–like automation, machine vision, and ever more numerous cameras and other sensors, making the face of any one of us a hyperlink to a vast database can soon happen without our knowing it.

4.   Automata, Augmentation, Identity

What do we call an augmented person? In the case of Stephen Hawking, "genius" usually suffices, even though, with his robotic wheelchair and voice synthesizer, he fits the definition of a cyborg pretty neatly.

What are the rules of the road for augmented humans in athletic competition? Should SAT proctors test for Ritalin among non-ADHD test takers? How will HR screeners evaluate "human+" job applicants?

From the machine end of the spectrum, what do we call a machine that can emulate human capabilities, sometimes uncannily? Alan Turing had one idea in 1950; many others have since been proposed.[3]

Binary distinctions will soon be overwhelmed by anomalies, a clear sign that today's simplistic typologies will fail.[4] Computing can approximate more and more human capabilities (such as understanding puns and riddles), while computers assume more and more mammalian forms, as Boston Dynamics' progress with legged robots illustrates. At the very least, the questions "What is a human?" and "What makes humans special?" will occupy more problematic terrain very soon: a 2015 article even proposed letting humans and robots marry.[5]

5. Humans Can Build Systems They Cannot Understand or Control

Perhaps the most vivid example of this tendency is the "flash crash" of 2010, in which the Dow Jones Industrial Average lost, then regained, 600 points in a matter of 20 minutes. It is widely believed that automated trading systems, overreacting to the gambit of a clever day trader in England, initially generated an excess of buy and sell orders, then retreated, causing a temporary lack of market liquidity.[6] If the aggregate wealth of the New York Stock Exchange can lose 9 percent of its value because of automated responses to manually generated spoof orders, given all the safeguards inherent in financial systems, what are the possibilities that millions of sensors could exhibit similar aberrant behavior? Code cannot easily be stress tested at this scale, nor can interactions among independent but interoperating systems be logically predicted. What are the rights and responsibilities of robot owners or robot makers?

Furthermore, the more we rely on machines to relieve ourselves of cognitive responsibilities, the more we forget how to do important things. An analysis of the 2009 crash of Air France flight 447, bound for Paris out of Rio de Janeiro, raises important questions about the erosion of human skills in the presence of automation: despite thousands of flight hours of experience, several crew members had extremely limited experience actually handling the aircraft, much less under challenging circumstances.[7] The U.S. Naval Academy dropped celestial navigation from the curriculum in 1997, given the far greater accuracy of GPS, although it still instructs cadets in the use of the sextant (but no longer with pad and paper).[8] Pocket calculators were most likely behind a generation of high school students not knowing how to add or subtract fractions, a necessity for carpenters and other tradespeople who have little use for decimal equivalents. Digital tools come with unintended consequences.

Three questions come to mind here. What are humans good at? What are computers good at? And how will human-computer partnerships change shape over the coming years?

Mind and Body

So we will be producing about 10^{26} to 10^{29} cps [calculations per second at global scale] of nonbiological computation per year in the early 2030s. This is roughly equal to our estimate for the capacity of all living biological human intelligence … . This state of computation in the early 2030s will not represent the Singularity, however, because it does not yet correspond to a profound expansion of our intelligence. By the

The more we rely on machines to relieve ourselves of cognitive responsibilities, the more we forget how to do important things.

mid-2040s, however, that one thousand dollars' worth of computation will be equal to 10^{26} cps, so the intelligence created per year (at a total cost of about \$1,012) will be about one billion times more powerful than all human intelligence today. That will indeed represent a profound change, and it is for that reason that I set the date for the Singularity—representing a profound and disruptive transformation in human capability—as 2045.

—Ray Kurzweil, *The Singularity Is Near*[9]

Kurzweil's singularity hypothesis—that machine cognitive capability will eclipse human capability with profound consequences—remains controversial. Indeed, the fact that Kurzweil is still a senior executive at Google, the world's leading robot company, raises some important questions.[10] Perhaps the most familiar critique of Kurzweil's thinking was made by Douglas Hofstadter, the Pulitzer Prize–winning author of *Gödel, Escher, Bach*. In a 2007 interview, Hofstadter summarized what many seem to have felt at the time: "If you read Ray Kurzweil's books and Hans Moravec's, what I find is that it's a very bizarre mixture of ideas that are solid and good with ideas that are crazy. It's as if you took a lot of very good food and some dog excrement and blended it all up so that you can't possibly figure out what's good or bad. It's an intimate mixture of rubbish and good ideas, and it's very hard to disentangle the two because these are smart people; they're not stupid."[11]

Anthony Damasio's brilliant book *Descartes' Error* posits a convincing alternative to Kurzweil's model of human cognition as a relatively simple and straightforward process that can be replicated and then surpassed in silicon. Rather than accept the Cartesian split of mind from body—expressed in the epigram "I think therefore I am" and central to Kurzweil's entire argument—Damasio reconnects thought with corporeality. The cognitive neuroscientist insists, with evidence, that it is emotion, the blurry juncture of mind *and* body, that enabled human survival in the evolutionary past and continues to define the species. All the talk about computer calculations equaling and surpassing human intelligence ignores this basic reality. Until computers can laugh, cry, sing, sweat nervously, and otherwise integrate mind and body, they cannot "surpass" what makes people people. Or, as Damasio explains, "I am not saying that the mind is in the body. I am saying that body contributes more than life support and modulatory effects to the brain. It contributes a *content* that is part and parcel of the workings of the normal mind."[12]

You don't have to be a neuroscientist to follow this argument. Our feelings often have a physical component: damp palms, tingling spine, quickened pulse and breathing. A central processing unit–like brain simply cannot account for such phenomena, let alone for muscle memory in athletes or perfect pitch in musicians. That said, algorithms and processing power and information storage and networking are absolutely increasing the power of nonhuman cognition. How will artificial intelligence/robotics incorporate these embodied forms of intelligence in as yet uninvented devices? AI poses potent questions, but maybe not the ones Kurzweil lists.

As I pointed out in the introduction with regard to Google+ and MIDI, technical defaults have long lives and broad effects. Indeed, the power of such defaults has been powerfully demonstrated by two researchers who found that countries with *opt-in* organ donor systems (such as the United States) have far lower levels of available transplant organs than do countries with *opt-out* systems.[13] We are getting to a point where such defaults are being set in regard to robotics, and they affect important human traits and processes.

Robots are probably best considered tools, as Asimov stated somewhat disingenuously once the field of robotics was established (his far more influential fiction implied otherwise). Humans and tools coevolve:[14] as we adapt to the many implications of our robots, the more we can self-consciously locate ourselves relative to them, the sooner we can design human-robot collaborations that can improve rather than impoverish our existence. And the more such self-awareness leads to explicit articulations rather than movie villains, literary tropes, or economic short-hand, the less mystery and confusion will attach to this new stage of technology innovation.

Humanity has always built tools, and tools always have unintended consequences. Those consequences have been substantial before: the rise of cities, the extension of human life spans, and the development of nuclear weapons. Before the next wave of change redefines work, caregiving, and warfare—even seeing and walking—it's time we have some straight talk about who we are relative to these machines, and what we expect from our transactions with them.

NOTES

Chapter 1

1. Bernard Roth, foreword to Bruno Siciliano and Oussama Khatib, eds., *Springer Handbook of Robotics* (Berlin: Springer-Verlag, 2008), viii.

2. See Matt McFarland, "Elon Musk: 'With Artificial Intelligence We Are Summoning the Demon,'" *The Washington Post* (blog), October 24, 2014, http://www.washingtonpost.com/blogs/innovations/wp/2014/10/24/elon-musk-with-artificial-intelligence-we-are-summoning-the-demon/.

3. An excellent history of AI, told by a participant, is Nils J. Nilsson, *The Quest for Artificial Intelligence: A History of Ideas and Achievements* (Cambridge: Cambridge University Press, 2010).

4. Ulrike Bruckenberger et al., "The Good, the Bad, the Weird: Audience Evaluation of a 'Real' Robot in Relation to Science Fiction and Mass Media," in G. Hermann et al., eds., *Social Robotics: 5th International Conference, ICSR 2013, Bristol, UK, October 27–29, 2013, Proceedings,* ICSR 2013, LNAI 8239, p. 301.

5. See W. Brian Arthur, *Increasing Returns and Path Dependence in the Economy* (Ann Arbor: University of Michigan Press, 1994), chapter 1.

6. This is a fascinating body of inquiry unto itself. For a compelling introduction, see Donald Norman, *The Design of Everyday Things* (1988; New York: Basic Books, 2002).

7. Jaron Lanier, *You Are Not a Gadget: A Manifesto* (New York: Knopf, 2010), 7–12.

8. Sergey Brin, as quoted in "Sergey Brin Live at Code Conference," *The Verge* (blog), May 27, 2014, http://live.theverge.com/sergey-brin-live-code-conference/.

9. Danny Palmer, "The future is here today: How GE is using the Internet of Things, big data and robotics to power its business," *Computing* 12 March 2015, http://www.computing.co.uk/ctg/feature/2399216/the-future-is-here-today-how-ge-is-using-the-internet-of-things-big-data-and-robotics-to-power-its-business/.

10. Chunka Mui and Paul B. Carroll, *Self-Driving Cars: Trillions Are Up for Grabs*, Kindle e-book (2013) location 223.

11. See Online Etymology Dictionary, "hello," http://www.etymonline.com/index.php?search=hello&searchmode=none/.

12. See Hugh Herr, "The New Bionics That Let Us Run, Climb, and Dance," *TED2014* (video blog), filmed March 2014, https://

www.ted.com/talks/hugh_herr_the_new_bionics_that_let_us_run_climb_and
_dance/.

13. See "Robin Millar: 'How Pioneering Eye Implant Helped My Sight,'" *BBC News* (blog), May 3, 2012, http://www.bbc.com/news/health-17936704/.

14. On solutionism, see Evgeny Morozov, *To Save Everything Click Here: The Folly of Technological Solutionism* (New York: Public Affairs, 2013), chapter 1.

15. A good starting point into this literature is Cass Sunstein and Richard Thaler, *Nudge: Improving Decisions about Health, Wealth, and Happiness* (New York: Penguin Books, 2008).

16. An essential reader on the topic is Patrick Lin, Keith Abney, and George A. Bekey, eds., *Robot Ethics: The Ethical and Social Implications of Robotics* (Cambridge, MA: MIT Press, 2012).

17. See Campaign to Stop Killer Robots, https://www.stopkillerrobots.org.

18. Steven Pinker, *How the Mind Works* (New York: Norton, 1999), 16.

19. Ray Kurzweil, *The Singularity Is Near: When Humans Transcend Biology* (New York: Viking, 2005), 4.

20. See Rodney Brooks, "Artificial Intelligence Is a Tool, Not a Threat," *Rethink Robotics* (blog), November 10, 2014, http://www.rethinkrobotics.com/blog/artificial-intelligence-tool-threat/.

Chapter 2

1. Illah Reza Nourbakhsh, *Robot Futures* (Cambridge, MA: MIT Press, 2013), xiv.

2. Rodney Brooks, *Flesh and Machines: How Robots Will Change Us* (Cambridge, MA: MIT Press, 2002), 13.

3. James L. Fuller, *Robotics: Introduction, Programming, and Projects* (Upper Saddle River, NJ: Prentice Hall, 1999), 3–4; emphasis added.

4. Cynthia Breazeal, *Designing Sociable Robots* (Cambridge, MA. MIT Press, 2004), 1.

5. Maja J. Mataric, *The Robotics Primer* (Cambridge, MA: MIT Press, 2007), 2.

6. Steve Kroft, "Are Robots Hurting Job Growth?" *60 Minutes* (video), January 13, 2013, http://www.cbsnews.com/video/watch/?id=50138922n/.

7. Vinton G. Cerf, "What's a Robot?" *Communications of the ACM* (Association for Computing Machinery) 56 (January 2013): 7; emphasis added.

8. George Bekey, *Autonomous Robots: From Biological Inspiration to Implementation and Control* (Cambridge, MA: MIT Press, 2005), 2; emphasis added.

9. My understanding of the duck tale relies on P. W. Singer, *Wired for War: The Robotics Revolution and Conflict in the 21st Century* (New York: Penguin Books, 2009), 42–43.

10. Isaac Asimov and Karen A. Frenkel, *Robots: Machines in Man's Image* (New York: Harmony Books, 1985), 13.

11. Asimov gives his editor John Campbell a great deal of credit for the structured formulation of the three laws; see *In Memory Yet Green: The Autobiography of Isaac Asimov 1920–1954* (Garden City, NY: Doubleday, 1979), 286.

12. Singer, *Wired for War*, 423.

13. Brooks, *Flesh and Machines*, 73.

14. Robin Murphy and David D. Woods, "Beyond Asimov: The Three Laws of Responsible Robotics," *IEEE Intelligent Systems* 24 (July–August 2009): 14–20, doi:10.1109/MIS.2009.69.

15. Joseph Engelberger, as quoted in Asimov and Frenkel, *Robots*, 25.

Chapter 3

1. Robert Geraci, *Apocalyptic AI: Visions of Heaven in Robotics, Artificial Intelligence, and Virtual Reality* (New York: Oxford University Press, 2010), 31.

2. Hiroaki Kitano, "The Design of the Humanoid Robot PINO," http://www.sbi.jp/symbio/people/tmatsui/pinodesign.htm, as quoted in Bekey, *Autonomous Robots*, 471.

3. See Hans P. Moravec, *Mind Children: The Future of Robot and Human Intelligence* (Cambridge, MA: Harvard University Press, 1988).

4. Geraci, *Apocalyptic AI*, 7.

5. Nourbakhsh, *Robot Futures*, 119.

6. Dwayne Day's plausible blog post suggests that the *Star Trek* writers borrowed from a White House pamphlet dating from 1958 that stated: "The first of these factors is the compelling urge of man to explore and to discover, the thrust of curiosity that leads *men to try to go where no one has gone before*." The piece also notes that Hollywood and the Southern California aerospace industry often cross-fertilized. See Dwayne A. Day, "Boldly Going: *Star Trek* and Spaceflight," *Space Review/Space News* (blog), November 28, 2005, http://www.thespacereview.com/article/506/1/.

7. See, for example, Leo Marx, *The Machine in the Garden: Technology and the Pastoral Ideal in America* (New York: Oxford University Press, 1965); Thomas P. Hughes, *American Genesis: A Century of Invention and Technological Enthusiasm* (New York: Viking, 1989); and David Nye, *America as Second Creation: Technology and Narratives of New Beginnings* (Cambridge, MA: MIT Press, 2003).

8. Evgeny Morozov, "The Perils of Perfection," *New York Times*, March 3, 2013, http://www.nytimes.com/2013/03/03/opinion/sunday/the-perils-of -perfection.html

9. William Edward Harkins, *Karel Čapek* (New York: Columbia University Press, 1962), 9.

10. Čapek quotation in London *Sunday Review*, as requoted in Karel Čapek, *R.U.R.* (New York: Pocket Books, 1973), reader's supplement, 11. "Rossum" was meant to connote logic, given that the Czech word "rozum" means "reason."

11. Čapek's dramatic work foreshadowed efforts to "teach" IBM's question-answering computer Watson how to play *Jeopardy!* roughly ninety years later by having it ingest *Wikipedia* and other online data repositories.

12. Čapek, *R.U.R.*, 49.

13. Ibid., 96.

14. Isaac Asimov, introduction to *The Complete Robot* (Garden City: Doubleday, 1982), xi.

15. Ibid., xii.

16. Norbert Wiener, *Cybernetics, or Communication and Control in the Animal and the Machine* (Cambridge, MA: MIT Press, 1948).

17. Phillip K. Dick, *Do Androids Dream of Electric Sheep?* (New York: Doubleday, 1968).

18. Pinker, *How the Mind Works*, 4.

19. The two essential English-language sources on Tezuka are Frederik L. Schodt, *The Astro Boy Essays: Osamu Tezuka, Mighty Atom, and the Manga/ Anime Revolution* (Berkeley, CA: Stone Bridge Press, 2007) and Helen McCarthy, *The Art of Osamu Tezuka: God of Manga* (New York: Abrams, 2009). I rely heavily on each of these in the following discussion.

20. See "20 Facts about Astro Boy," *Geordie Japan: A Guide to Finding Japan in Newcastle-upon-Tyne* (blog), January 10, 2013, http://geordiejapan .wordpress.com/2013/01/10/20-facts-about-astro-boy/.

21. Reprinted from Schodt's translation of the Japanese in *The Astro Boy Essays*, 108.

Chapter 4
1. See "Global Industrial Robot Sales Rose 27 [Percent] in 2014," *Reuters*, March 22, 2015, http://www.reuters.com/article/industry-robots-sales -idUSL6N0WM1NS20150322/.

2. See "Foxconn to Rely More on Robots; Could Use 1 Million in 3 years," *Reuters*, August 1, 2011, http://www.reuters.com/article/us-foxconn-robots -idUSTRE77016B20110801/.

3. For more on robot locomotion, see Roland Siegwart and Illah R. Nourbakhsh, *Introduction to Autonomous Mobile Robots* (Cambridge, MA: MIT Press, 2004), chapter 2.

4. Singer, *Wired for War*, 55.

5. Nourbakhsh, *Robot Futures*, 49–50.

6. Municipalities are buying large numbers of license-plate cameras, which rapidly pay for themselves by identifying cars with expired licenses and outstanding parking tickets, or stolen vehicles. A typical system can scan more than 750 cars an hour. See Shawn Musgrave, "Big Brother or Better Police Work? New Technology Automatically Runs License Plates ... of Everyone," *Boston Globe*, April 8, 2013.

7. Bekey, *Autonomous Robots*, 104–7; Brooks, *Flesh and Machines*, 36–43.

8. Brooks, *Flesh and Machines*, 72–73.

9. Siegwart and Nourbakhsh, *Introduction to Autonomous Mobile Robots*, chapter 6.

10. The quadrotors (quadcopters) at the University of Pennsylvania GRASP lab are an example of robots deployed in groups. See https:// www.grasp.upenn.edu.

11. Bekey, *Autonomous Robots*, 5–6.

12. Singer, *Wired for War*, 60.

13. See "Military Robot Markets to Exceed $8 Billion in 2016," *ABIresearch: Intelligence for Innovators* (blog), February 15, 2011, http://www.abiresearch.com/ press/military-robot-markets-to-exceed-8-billion-in-2016/.

14. See Cloud Robotics and Automation, http://goldberg.berkeley.edu/ cloud-robotics/.

15. See RoboCup, http://www.robocup.org/about-robocup/objective/.

Chapter 5

1. Mui and Carroll, *Driverless Cars*, location 13.

2. Sebastian Thrun, "Toward Robotic Cars," *Communications of the ACM* 53 (April 2010): 99; and Mui and Carroll, *Driverless Cars*, location 43.

3. Thrun, "Toward Robotic Cars."

4. Leo Kelion, "Audi Claims Self-Drive Car Speed Record after German Test," *BBC News* (blog), October 21, 2014, http://www.bbc.com/news/ technology-29706473/.

5. Casey Newton, "Uber Will Eventually Replace All Its Drivers with Self-Driving Cars, *The Verge* (blog), May 28, 2014, http://www.theverge.com/2014/5/28/5758734/uber-will-eventually-replace-all-its-drivers-with-self-driving-cars/.

6. Douglas Macmillan, "GM Invests $500 Million in Lyft, Plans System for Self-Driving Cars: Auto Maker Will Work to Develop System That Could Make Autonomous Cars Appear at Customers' Doors," *Wall Street Journal*, January 4, 2016, http://www.wsj.com/articles/gm-invests-500-million-in-lyft-plans-system-for-self-driving-cars-1451914204/.

7. See Shaun Bailey, "BMW Track Trainer: How a Car Can Teach You to Drive," *Road & Track* (blog), September 7, 2011, http://www.roadandtrack.com/car-culture/a17638/bmw-track-trainer/.

8. Frank Levy and Richard Murnane, *The New Division of Labor: How Computers Are Creating the Next Job Market* (New York: Russell Sage Foundation; Princeton: Princeton University Press, 2004), 20.

9. Defense Advanced Research Projects Agency (DARPA), "Report to Congress: DARPA Prize Authority: Fiscal Year 2005 Report in Accordance with U.S.C. §2374a," released March 2006, 3, http://archive.darpa.mil/grandchallenge/docs/Grand_Challenge_2005_Report_to_Congress.pdf.

10. See Erico Guizzo, "How Google's Self-Driving Car Works," *IEEE Spectrum*, October 18, 2011, http://spectrum.ieee.org/automaton/robotics/artificial-intelligence/how-google-self-driving-car-works/.

11. See Alex Davies, "This Palm-Sized Laser Could Make Self-Driving Cars Way Cheaper," *Wired* (blog), September 25, 2014, http://www.wired.com/2014/09/velodyne-lidar-self-driving-cars/.

12. Sebastian Thrun et al., "Stanley: The Robot that Won the DARPA Grand Challenge," *Journal of Field Robotics* 23 (2009): 665.

13. See "What If It Could Be Easier and Safer for Everyone to Get Around?" *Google Self-Driving Project* (video/text blog) [no date], https://www.google.com/selfdrivingcar/.

14. *Car and Driver*, August 2013, cover.

15. See James Vincent, "Toyota's $1 Billion AI Company Will Develop Self-Driving Cars and Robot Helpers," *The Verge* (blog), November 6, 2015, http://www.theverge.com/2015/11/6/9680128/toyota-ai-research-one-billion-funding/.

16. See Nic Fleming and Daniel Boffey, "Lasers-Guided Cars Could Allow Drivers to Eat and Sleep at the Wheel While Travelling in 70 mph Convoys," *Daily Mail.com* (blog), June 22, 2009, http://www.dailymail.co.uk/

sciencetech/article-1194481/Lasers-guided-cars-allow-eat-sleep-wheel
-travelling-70mph-convoys.html

17. See Brad Templeton, "I Was Promised Flying Cars!" *Templetons.com* (blog)
[no date], http://www.templetons.com/brad/robocars/roadblocks.html

18. See Daniel Kahneman, *Thinking, Fast and Slow* (New York, Farrar, Straus
and Giroux, 2011), chapters 12 and 13.

19. See Bruce Schneier, "Virginia Tech Lesson: Rare Risks Breed Irrational
Responses," *Wired* (blog), May, 2007, https://www.schneier.com/essays/
archives/2007/05/virginia_tech_lesson.html

20. Zack Rosenberg, "The Autonomous Automobile," *Car and Driver*,
August 2013, 68, http://www.caranddriver.com/features/the-autonomous
-automobile-the-path-to-driverless-cars-explored-feature/.

21. See Chris Urmson, "The View from the Front Seat of the Google
Self-Driving Car, Chapter 2," *Medium.com* (blog), July 16, 2015, https://
medium.com/@chris_urmson/the-view-from-the-front-seat-of-the-google-
self-driving-car-chapter-2-8d5e2990101b#.17cg8dyt4.

22. See Lee Gomes, "Driving in Circles: The Autonomous Google Car May
Never Actually Happen," *Slate* (blog), October 21, 2014, http://www.slate.com/
articles/technology/technology/2014/10/google_self_driving_car_it_may
_never_actually_happen.html

23. See Nick Bilton, "The Money Side of Driverless Cars," *The New York
Times* (blog), July 9, 2013, http://bits.blogs.nytimes.com/2013/07/09/
the-end-of-parking-tickets-drivers-and-car-insurance/.

24. See Shawna Ohm, "Why UPS Drivers Don't Make Left Turns," *Yahoo! Fi-
nance* (video/text blog), September 30, 2014, http://finance.yahoo.com/news/
why-ups-drivers-don-t-make-left-turns-172032872.html

25. For more about how much we spend on cars and on what, see Mui and
Carroll, *Self-Driving Cars,* chapter 1.

26. Ibid., location 127.

27. See Centers for Disease Control/National Center for Health Statistics,
"FastStats: Accidental or Unintentional Injuries," last updated September 30,
2015, http://www.cdc.gov/nchs/fastats/accidental-injury.htm.

28. See Centers for Disease Control, "National Hospital Ambulatory Medical Care Survey: 2010 Emergency Department Survey Tables," http://
www.cdc.gov/nchs/data/ahcd/nhamcs_emergency/2010_ed_web_tables.pdf.

29. Mui and Carroll, *Self-Driving Cars*, location 74.

30. See Climateer, "Understanding the Future of Mobility: On-Demand
Driverless Cars," *Climateer Investing* (blog), August 10, 2015, http://

climateerinvest.blogspot.co.uk/2015/08/understanding-future-of-mobility
-on.html

31. See U.S. Public Interest Research Group and Frontier Group, "Transportation and the New Generation: Why Young People Are Driving Less and What It Means for Transportation Policy" (report), released April 5, 2012, http://www.uspirg.org/reports/usp/transportation-and-new-generation/.

32. See Mark Strassman, "A Dying Breed: The American Shopping Mall," *CBS News.com* (video/text blog), March 23, 2014, http://www.cbsnews.com/news/a-dying-breed-the-american-shopping-mall/.

33. See, for example, Laura Houston Santhanam, Amy Mitchell, and Tom Rosenstiel, "The State of the News Media 2012: An Annual Report," Pew Research Center's Project for Excellence in Journalism, http://stateofthemedia.org/2012/audio-how-far-will-digital-go/audio-by-the-numbers/.

34. See Steve Mahan, "Self-Driving Car Test," YouTube.com (video), March 28, 2012, https://www.youtube.com/watch?v=cdgQpa1pUUE/.

35. See Lucia Huntington, "The Real Distraction at the Wheel: Texting Is a Big Problem, but with More People Eating and Driving Than Ever Before, Maybe That's an Even Bigger Problem," *The Boston Globe* (blog), October 14, 2009, http://www.boston.com/lifestyle/food/articles/2009/10/14/dining_while_driving_theres_many_a_slip_twixt_cup_and_lip_but_that_doesnt_stop_us/.

36. See William H. Janeway, "From Atoms to Bits to Atoms: Friction on the Path to the Digital Future," *Forbes.com* (blog), July 30, 2015, http://www.forbes.com/sites/valleyvoices/2015/07/30/from-atoms-to-bits-to-atoms-friction-on-the-path-to-the-digital-future/.

37. See Erin Griffith, "If Driverless Cars Save Lives, Where Will We Get Organs?" *Fortune* (blog), August 15, 2014, http://fortune.com/2014/08/15/if-driverless-cars-save-lives-where-will-we-get-organs/.

38. See Alan S. Blinder, "Offshoring: The Next Industrial Revolution?" *Foreign Affairs*, March–April 2006, http://www.foreignaffairs.com/articles/61514/alan-s-blinder/offshoring-the-next-industrial-revolution/.

39. See Megahn Walsh, "Why No One Wants to Drive a Truck Anymore: Commercial Drivers' Average Age is 55, and Young People Don't Want to Take Up the Slack," *BloombergBusiness* (blog), November 14, 2013, http://www.bloomberg.com/news/articles/2013-11-14/2014-outlook-truck-driver-shortage/.

40. See Adario Strange, "Mercedes-Benz Unveils Self-Driving 'Future Truck' on Germany's Autobahn," *Mashable* (video/text blog), July 6, 2014, http://mashable.com/2014/07/06/mercedes-benz-self-driving-truck/.

41. RAND study, as quoted in Mui and Carroll, *Self-Driving Cars*, location 279.

42. Mui and Carroll, *Self-Driving Cars*, location 214.

Chapter 6

1. See DARPA, "Mission," http://www.darpa.mil/about-us/mission/.

2. See "Beyond the Borders of 'Possible,'" *army.mil*, January 27, 2015 (interview with Dr. Bradford Tousley, director of DARPA's Tactical Technology Office or TTO, by staff of U.S. Army's *Access AL&T* magazine), http://www.army.mil/mobile/article/?p=141732/.

3. Ronald C. Arkin, *Governing Lethal Behavior in Autonomous Robots* (New York: CRC Press, 2009), xii.

4. See Jeremiah Gertler, "U.S. Unmanned Aerial Systems," Congressional Research Service report R42136, January 3, 2012, http://www.fas.org/sgp/crs/natsec/R42136.pdf.

5. Singer, *Wired for War*, 33.

6. Ibid., 36.

7. See R. Jeffrey Smith, "High-Priced F-22 Fighter Has Major Shortcomings," *Washington Post*, July 10, 2009, http://www.washingtonpost.com/wp-dyn/content/article/2009/07/09/AR2009070903020.html?hpid=topnews&sub=AR&sid=ST2009071001019/.

8. See Brian Bennett, "Predator Drones Have Yet to Prove Their Worth on Border," *Los Angeles Times*, April 28, 2012, http://articles.latimes.com/2012/apr/28/nation/la-na-drone-bust-20120429.

9. See Singer, *Wired for War*, 114–16.

10. See "Autonomous Underwater Vehicle—Seaglider," *kongsberg.com* [no date], http://www.km.kongsberg.com/ks/web/nokbg0240.nsf/AllWeb/EC2FF8B58CA491A4C1257B870048C78C?OpenDocument/.

11. See AUVAC (Autonomous Undersea Vehicle Applications Center), "AUV System Spec Sheet: Proteus Configuration," *auvac.org* [no date], http://auvac.org/configurations/view/239/.

12. Singer, *Wired for War*, 114–15.

13. See Rafael Advanced Defense Systems, Ltd., "Protector Unmanned Naval Patrol Vehicle," rafael.co.il [no date], http://www.rafael.co.il/Marketing/351-1037-en/Marketing.aspx.

14. See "iRobot Delivers 3,000th PackBot," investor.irobot.com (news release), February 16, 2010, http://investor.irobot.com/phoenix.zhtml?c=193096&p=irol-newsArticle&ID=1391248/.

15. See QinetiQ North America, "TALON® Robots: From Reconnaissance to Rescue, Always Ready on Any Terrain," *QinetiQ-NA.com* (data sheet) [no date], https://www.qinetiq-na.com/wp-content/uploads/data-sheet_talon.pdf.

16. "March of the Robots," *Economist*, June 2, 2012, http://www.economist.com/node/21556103/.

17. See Evan Ackerman and Erico Guizzo, "DARPA Robotics Challenge: Amazing Moments, Lessons Learned, and What's Next," *IEEE Spectrum*, June 11, 2015, http://spectrum.ieee.org/automaton/robotics/humanoids/darpa-robotics-challenge-amazing-moments-lessons-learned-whats-next/.

18. See Sydney J. Freedberg Jr., "Why the Military Wants Robots with Legs (Not to Run Faster Than Usain Bolt)," *Breaking Defense* (blog), September 7, 2012. http://breakingdefense.com/2012/09/07/why-the-military-wants-robots-with-legs-robot-runs-faster-than/.

19. Ronald C. Arkin, "Ethical Robots in Warfare," *IEEE Technology and Society Magazine* 28 (Spring 2009), http://www.dtic.mil/dtic/tr/fulltext/u2/a493429.pdf.

20. See Human Rights Watch, "The 'Killer Robots' Accountability Gap," *hrw.org* (blog), April 8, 2015, https://www.hrw.org/news/2015/04/08/killer-robots-accountability-gap/.

21. See UN General Assembly, Human Rights Council, "Report of the Special Rapporteur on Extrajudicial, Summary or Arbitrary Executions, Christof Heyns," A/HRC/23/47, April 17, 2013, http://www.ohchr.org/Documents/HRBodies/HRCouncil/RegularSession/Session23/A.HRC.23.47_EN.pdf.

22. Many of these points echo Arkin, "Ethical Robots in Warfare."

23. See Associated Press, "Afghan Panel: U.S. Airstrike Killed 47 in Wedding Party," *Washington Post*, July 12, 2008, http://articles.washingtonpost.com/2008-07-12/world/36906336_1_civilians-airstrike-afghan-panel/.

24. See David S. Cloud, "Civilian Contractors Playing Key Roles in U.S. Drone Operations," *Los Angeles Times*, December 29, 2011, http://articles.latimes.com/2011/dec/29/world/la-fg-drones-civilians-20111230/.

25. See, for example, Human Rights Watch, "Losing Humanity: The Case against Killer Robots," November 2012, especially sections II and III http://www.hrw.org/sites/default/files/reports/arms1112ForUpload_0_0.pdf.

26. See "*Dr. Strangelove, or: How I Learned to Stop Worrying and Love the Bomb*: Plot Summary," *IMDb.com* [no date] http://www.imdb.com/title/tt0057012/plotsummary?ref_=tt_stry_pl/.

27. See Christopher Mims, "U.S. Military Chips 'Compromised,'" *MIT Technology Review*, May 30, 2012, http://www.technologyreview.com/view/428029/us-military-chips-compromised/.

28. See "Interview with Defense Expert P. W. Singer: 'The Soldiers Call It War Porn,'" *Spiegel Online International*, March 12, 2010, http://www.spiegel.de/international/world/interview-with-defense-expert-p-w-singer-the-soldiers-call-it-war-porn-a-682852.html

29. See *New York Times*, "Distance from Carnage Doesn't Prevent PTSD for Drone Pilots," atwar.nytimes.com (blog), February 23, 2013, http://atwar.blogs.nytimes.com/2013/02/25/distance-from-carnage-doesnt-prevent-ptsd-for-drone-pilots/ and Christopher Drewand Dave Philipps, "As Stress Drives Off Drone Operators, Air Force Must Cut Flights," *New York Times*, June 16, 2015, http://www.nytimes.com/2015/06/17/us/as-stress-drives-off-drone-operators-air-force-must-cut-flights.html

30. See Chris Woods, "Drone Warfare: Life on the New Frontline," *The Guardian*, February 24, 2015, http://www.theguardian.com/world/2015/feb/24/drone-warfare-life-on-the-new-frontline/.

31. Mubashar Jawed Akbar, as quoted in Singer, *Wired for War*, 312.

32. Singer, *Wired for War*, 198.

Chapter 7

1. On ATMs, see John M. Jordan, *Information, Technology, and Innovation: Resources for Growth in a Connected World* (Hoboken, NJ: Wiley, 2012), 153–55.

2. Erik Brynjolfsson and Andrew McAfee, *Race against the Machine: How the Digital Revolution Is Accelerating Innovation, Driving Productivity, and Irreversibly Transforming Employment and the Economy* (Lexington, MA: Digital Frontier Press, 2011), Kindle edition. Much of the material from this e-book appears in Brynjolfsson and McAfee's more comprehensive print book, *The Second Machine Age: Work, Progress, and Prosperity in a Time of Brilliant Technologies* (New York: Norton, 2014).

3. See David Autor, "The 'Task' Approach to Labor Markets: An Overview," National Bureau of Economic Research Working Paper 18711, http://www.nber.org/papers/w18711/.

4. See also Levy and Murnane, *The New Division of Labor*, 6. Levy and Murnane have also coauthored papers with Autor.

5. Autor, "The 'Task' Approach to Labor Markets," 5.

6. See IFR (International Federation of Robotics), "Industrial Robot Statistics," in "World Robotics 2015 Industrial Robots," *ifr.org* (report) [no date],http://www.ifr.org/industrial-robots/statistics/.

7. See Sam Grobart, "Robot Workers: Coexistence Is Possible," *Bloomberg-Business* (blog), December 13, 2012, http://www.bloomberg.com/news/articles/2012-12-13/robot-workers-coexistence-is-possible/.

8. "But what they do have is software that makes sure the robots are in the right place at the right time. This was a software play." Jim Tompkins, as quoted in Sam Grobart, "Amazon's Robotic Future: A Work in Progress, *BloombergBusiness* (blog), November 30, 2012, http://www.bloomberg.com/news/articles/2012-11-30/amazons-robotic-future-a-work-in-progress/.

9. See Kevin Bullis, "Random-Access Warehouses: A Company Called Kiva Systems Is Speeding Up Internet Orders with Robotic Systems That Are Modeled on Random-Access Computer Memory," *MIT Technology Review*, November 8, 2007, http://www.technologyreview.com/news/409020/random-access-warehouses/.

10. Robert B. Reich, *The Work of Nations: Preparing Ourselves for 21st Century Capitalism* (New York: Vintage, 1992).

11. See "The Age of Smart Machines: Brain Work May Be Going the Way of Manual Work," *Economist*, My 23, 2013, http://www.economist.com/news/business/21578360-brain-work-may-be-going-way-manual-work-age-smart-machines/.

12. John Markoff, "Armies of Expensive Lawyers, Replaced by Cheaper Software," *New York Times*, March 4, 2011, http://www.nytimes.com/2011/03/05/science/05legal.html?pagewanted=all/.

13. U.S. Census Bureau, "Historical Income Tables: Households," [no date], http://www.census.gov/hhes/www/income/data/historical/household/index.html

14. For one example, see Thomas Hungerford, "Changes in Income Inequality among U.S. Tax Filers between 1991 and 2006: The Role of Wages, Capital Income, and Taxes," Economic Policy Institute working paper, January 23, 2013, http://papers.ssrn.com/sol3/papers.cfm?abstract_id=2207372/.

15. U.S. Bureau of Labor Statistics, graph of productivity and average real earnings against index relative to 1970, from about 1947 to 2009, https://thecurrentmoment.files.wordpress.com/2011/08/productivity-and-real-wages.jpg.

16. See Illah Nourbakhsh, "Will Robots Boost Middle-Class Unemployment?" *Quartz*, June 7, 2013, http://qz.com/91815/the-burgeoning-middle-class-of-robots-will-leave-us-all-jobless-if-we-let-it/.

17. Gill Pratt, "Robots to the Rescue," *Bulletin of the Atomic Scientists*, December 3, 2013. http://thebulletin.org/robot-rescue/.

18. See Kevin Kelly, "Better Than Human: Why Robots Will—and Must—Take Our Jobs," *Wired*, December 24, 2012, http://www.wired.com/gadgetlab/2012/12/ff-robots-will-take-our-jobs/.

19. See Steven Cherry, "Robots Are Not Killing Jobs, Says a Roboticist: A Georgia Tech Professor of Robotics Argues Automation Is Still Creating More Jobs Than It Destroys," *IEEE Spectrum*, April 9, 2013, http://spectrum.ieee.org/podcast/robotics/industrial-robots/robots-are-not-killing-jobs-says-a-roboticist/.

20. Levy and Murnane, *The New Division of Labor*, 2.

21. See James Bessen, "Employers Aren't Just Whining—The 'Skills Gap' Is Real," *Harvard Business Review*, August 25, 2014, https://hbr.org/2014/08/employers-arent-just-whining-the-skills-gap-is-real/.

22. See David Wessel, "Software Raises Bar for Hiring," *Wall Street Journal*, May 31, 2012, http://www.wsj.com/articles/SB10001424052702304821304577436172660988042/.

23. For more on disability, see the 2013 NPR package by Chana Joffe-Walt entitled "Unfit for Work: The Startling Rise of Disability in America," http://apps.npr.org/unfit-for-work/.

24. Ibid.

Chapter 8

1. Robin R. Murphy and Debra Schreckenghost, "Survey of Metrics for Human-Robot Interaction," *HRI 2013 Proceedings: 8th ACM/IEEE International Conference on Human-Robot Interaction*, 197.

2. Ibid.

3. Cynthia Breazeal, Atsuo Takanashi, and Tetsunori Kobayashi, "Social Robots That Interact with People," in Siciliano and Khatib, *Springer Handbook of Robotics*, 1349–50.

4. My discussion in this section relies heavily on Robin R. Murphy et al., "Search and Rescue Robotics," in Siciliano and Khatib, *Springer Handbook of Robotics*, 1151–73.

5. See ibid., 1173n42.

6. See Lawrence Diller, MD, "The NFL's ADHD, Adderall Mess," *The Huffington Post* (blog), February 5, 2013, http://www.huffingtonpost.com/news/NFL+Suspensions/.

7. See Ashlee Vance, "Dinner and a Robot: My Night Out with a PR3," *BloombergBusiness*, August 9, 2012, http://www.bloomberg.com/news/articles/2012-08-09/dinner-and-a-robot-my-night-out-with-a-pr2#r=lr-fst/.

8. Intuitive Surgical 2014 annual report, p. 45 http://www.annualreports
.com/Company/intuitive-surgical-inc/.

9. See Herb Greenberg, "Robotic Surgery: Growing Sales, but Growing
Concerns," *CNBC*, March 19, 2013, http://www.cnbc.com/id/100564517/;
and Roni Caryn Rabin, "Salesmen in the Surgical Suite," *New York Times*,
March 25, 2013, http://www.nytimes.com/2013/03/26/health/salesmen-in
-the-surgical-suite.html?pagewanted=all/.

10. See Citron Research, "Intuitive Surgical: Angel with Broken Wings, or
Devil in Disguise?" (report), January 17, 2013, http://www.citronresearch
.com/wp-content/uploads/2013/01/Intuitive-Surgical-part-two-final.pdf;
and Lawrence Diller, MD, et al., "Robotically Assisted vs. Laparascopic Hys-
terectomies among Women with Benign Gynecological Disease," *JAMA:
Journal of the American Medical Association* 309 (February 20, 2013), http://
jama.jamanetwork.com/article.aspx?articleid=1653522/.

11. See Ceci Connolly, "U.S. Combat Fatality Rate Lowest Ever: Technology
and Surgical Care at the Front Lines Credited with Saving Lives," *Washington Post*, December 9, 2004, A26, http://www.washingtonpost.com/wp-dyn/
articles/A49566-2004Dec8.html

12. See Nitish Thakor, "Building Brain Machine Interfaces—Neuro-
prosthetic Control with Electrocorticographic Signals," *IEEE Lifesciences*,
April 2012, http://lifesciences.ieee.org/publications/newsletter/april-2012/
96-building-brain-machine-interfaces-neuroprosthetic-control-with
-electrocorticographic-signals/.

13. See "How Would You Like Your Assistant—Human or Robotic?" *Georgia Tech News Center*, April 29, 2013, http://www.gatech.edu/newsroom/
release.html?nid=210041/.

14. See "Home Care Robot, 'Yurina,'" *DigInfoTV* (video/text), August 12,
2010, http://www.diginfo.tv/v/10-0137-r-en.php.

15. See SECOM, "Meal-Assistance Robot My Spoon Allows Eating with Only
Minimal Help from a Caregiver," seco.co.jp [no date], http://www.secom.co.jp/
english/myspoon/.

16. See Miwa Suzuki, "'Welfare Robots' to Ease Burden in Greying Japan,"
Phys.org, July 29, 2010, http://phys.org/news/2010-07-welfare-robots-ease-
burden-greying.html

17. See Anne Tergesen and Miho Inada, "It's Not a Stuffed Animal, It's a
$6,000 Medical Device: Paro the Robo-Seal Aims to Comfort the Elderly, but Is
It Ethical?" *Wall Street Journal*, June 21, 2010, http://online.wsj.com/article/
SB10001424052748704463504575301051844937276.html?

18. See, for example, Sherry Turkle, *Alone Together: Why We Expect More from
Technology and Less from Each Other* (New York: Basic Books, 2012).

19. Amanda Sharkey and Noel Sharkey, "Granny and the Robots: Ethical Issues in Robot Care for the Elderly," *Ethics of Information Technology* 14 (2012): 35.

20. Bekey, *Autonomous Robots*, 512.

21. Brynjolfsson and McAfee, *Second Machine Age,* 96.

22. See "Pros Rake in More Chips Than Computer Program during Poker Contest, but Scientifically Speaking, Human Lead Not Large Enough to Avoid Statistical Tie," *Carnegie Mellon University News*, May 8, 2015, http://www.cmu.edu/news/stories/archives/2015/may/poker-pros-rake-in-more-chips.html

23. See Mark Prigg, "Robots Take the Checquered Flag: Watch the Self Driving Racing Car That Can Beat a Human Driver," *Daily Mail*, March 20, 2016, http://www.dailymail.co.uk/sciencetech/article-2959134/Robots-chequered-flag-Watch-self-driving-racing-car-beat-human-driver-sometimes.html

24. See "Computational Aesthetics Algorithm Spots Beauty That Humans Overlook: Beautiful Images Are Not Always Popular Ones, Which Is Where the Crowd Beauty Algorithm Can Help, Say Computer Scientists," *MIT Technology Review*, May 22, 2015, http://www.technologyreview.com/view/537741/computational-aesthetics-algorithm-spots-beauty-that-humans-overlook/.

25. See "*The Economist* Explains How Machine Learning Works," *The Economist* (blog), May 13, 2015, http://www.economist.com/blogs/economist-explains/2015/05/economist-explains-14/.

26. See "Exploring the Epic Chess Match of Our Time," *FiveThirtyEight* (video/text), October 22, 2014, http://fivethirtyeight.com/features/the-man-vs-the-machine-fivethirtyeight-films-signals/.

27. See Tyler Cowen, "What are humans still good for? The turning point in Freestyle chess may be approaching," *Marginal Revolution: Small Steps toward a Much Better World*, November 5, 2013, http://marginalrevolution.com/marginalrevolution/2013/11/what-are-humans-still-good-for-the-turning-point-in-freestyle-chess-may-be-approaching.html and Mike Cassidy, "Centaur Chess Brings Out the Best in Humans and Machines," *Bloom-Research* (blog), December 14, 2014, http://bloomreach.com/2014/12/centaur-chess-brings-best-humans-machines/.

28. See Walter Frick, "When Your Boss Wears Metal Pants," *Harvard Business Review*, June 2015, https://hbr.org/2015/06/when-your-boss-wears-metal-pants/.

29. See Lindsay Fortago, Philip Stafford, and Aliya Ram, "Flash Crash: Ten Days in Hounslow," *Financial Times*, April 22, 2015, http://www.ft.com/intl/cms/s/0/9d7e50a4-e906-11e4-b7e8-00144feab7de.html#axzz43Y1pFxDA/.

30. Byron Reeves and Clifford Nass, *The Media Equation: How People Treat Computers, Television, and New Media Like People and Places* (New York: Cambridge University Press, 1996), 4.

31. Turkle, *Alone Together*.

32. Sherry Turkle, *Life on the Screen*, as quoted in Brooks, *Flesh and Machines*, 149.

33. See "iRobot's PackBot on the Front Lines," *Phys.org*, February 24, 2006, http://phys.org/news11166.html#jCp.Phys.org.

34. Singer, *Wired for War*, 338.

35. See M. K. Lee et al., "Ripple Effects of an Embedded Social Agent: A Field Study of a Social Robot in the Workplace," in *Proceedings of CHI 2012*, http://www.cs.cmu.edu/~kiesler/publications/2012/Ripple-Effects-Embedded-Agent-Social-Robot.pdf.

36. Brooks, *Flesh and Machines*, 180.

37. Ibid., 236.

Chapter 9

1. Brynjolfsson and McAfee, *Second Machine Age*, 159–62.

2. On intergenerational mobility, see Tony Judt, *Ill Fares the Land* (New York: Penguin Books, 2010).

3. Jacob Aron, "Forget the Turing test—there are better ways of judging AI," *New Scientist*, September 21, 2015, https://www.newscientist.com/article/dn28206-forget-the-turing-test-there-are-better-ways-of-judging-ai/.

4. This is, of course, the path of a "paradigm shift" posited by Thomas Kuhn in *The Structure of Scientific Revolutions* (Chicago: University of Chicago Press, 1962).

5. See Gary Marchant, "A.I. Thee Wed: Humans Should Be Able to Marry Robots," *Slate*, August 10, 2015, http://www.slate.com/articles/technology/future_tense/2015/08/humans_should_be_able_to_marry_robots.html

6. See Securities and Exchange Commission (SEC), "Findings Regarding the Market Events of May 6, 2010: Report of the Staffs of the CFTC and SEC to the Joint Advisory Committee on Emerging Regulatory Issues," September 30, 2010, http://www.sec.gov/news/studies/2010/marketevents-report.pdf

7. See William Langewiesche, "The Human Factor," *Vanity Fair* September 17, 2014, http://www.vanityfair.com/news/business/2014/10/air-france-flight-447-crash/.

8. See "Despite Buzz, Navy Will Still Teach Stars," *Ocean Navigator*, January–February 2003, http://www.oceannavigator.com/January-February-2003/Despite-buzz-Navy-will-still-teach-stars/.

9. Kurzweil, *The Singularity Is Near*, 135–36.

10. As of 2016, Google shuffled the leadership of the robotics team. To be clear, I am asserting not that Kurzweil is at all connected with management of the Google (or Alphabet) robotics efforts, rather that his hiring by the same company could reflect a corporate ethos or commitment that links the Singularity and robot commercialization. See Connor Dougherty, "Alphabet Shakes Up Its Robotics Division," *New York Times*, January 15, 2016, http://www.nytimes.com/2016/01/16/technology/alphabet-shakes-up-its-robotics-division.html

11. See Greg Ross, "Interview with Douglas Hofstadter" (conducted January 2007), *American Scientist* [no date], http://www.americanscientist.org/bookshelf/pub/douglas-r-hofstadter/.

12. Antonio Damasio, *Descartes' Error: Emotion, Reason, and the Human Brain* (New York: Putnam, 1994), 226.

13. H. P. van Dalen and K. Henkens, "Comparing the Effects of Defaults in Organ Donation Systems," *Social Science and Medicine* 106 (2014): 137–42.

14. See, for example, Frank Geels, "Co-Evolution of Technology and Society: The Transition in Water Supply and Personal Hygiene in the Netherlands (1850–1930)—a Case Study in Multi-Level Perspective," *Technology in Society* 27 (2005): 363–97.

GLOSSARY

AGV (Automated Guided Vehicle)
Unmanned sled- or truck-like wheeled vehicle that follows a preprogrammed path in a facility, often for delivery of routine supplies. Unlike a robotic vehicle, an AGV is neither autonomous nor remotely controlled.

AI (Artificial Intelligence)
Branch of computer science that concerns itself with the computerized re-creation of human cognition, either generally or in a delimited domain; any such re-creation. Machine vision and machine learning (including speech recognition) are two subbranches of AI that bear on robotics.

Android
Traditionally, "an automaton resembling a human being" (*Oxford English Dictionary*).

Big Data
Data sets whose size exceeds the capabilities of traditional data processing. Because the data volumes generated by masses of robots, each of which has a suite of sensors, can be substantial, big data tools are frequently enlisted to manage and analyze them.

Cyborg
Being that merges artificial and organic systems of control. In the context of robotics, a cyborg is typically an augmented human who has been enhanced through the application of computational or robotic capabilities.

DARPA (Defense Advanced Research Projects Agency)
Agency of the U.S. Department of Defense responsible for the development of emerging technologies for use by the military. DARPA has aggressively supported autonomous vehicle and robotic research.

Human-Robot Interaction (HRI)
Relatively understudied branch of robotics, particularly with regard to humans' reactions to autonomous robots in their midst.

Lidar
Remote sensing technology that measures distance by illuminating a target with a laser and analyzing the reflected light. Lidar was a crucial component on the first generation of Google's self-driving cars.

Moore's Law
Intel cofounder Gordon Moore's (1965) observation that the number of transistors on an integrated circuit and thus also its overall processing power double roughly every two years (they have done so for more than fifty years). Since many robot tasks are computationally intensive, increased processor power makes more of these tasks feasible or cost effective.

Path Dependence
In the technological domain, the notion that current options are constrained by past decisions: railroad gauges, typewriter layouts, and word-processing software are common examples of how path dependence may prevent a superior innovation from winning out in the marketplace.

Robot
According to pioneer roboticist George Bekey, a "*machine that senses, thinks, and acts.* Thus, a robot must have sensors, processing ability that emulates some aspects of cognition, and actuators." Culturally speaking, a robot tends to be a mechanical entity that exhibits humanlike capabilities.

Robotics
Disciplines that combine to study, design, and build robots: with computer science in the forefront, robotics draws also on materials science, psychology, statistics, mathematics, physics, and engineering. The term was coined by science fiction writer Isaac Asimov in the 1940s.

Sensors
Sensing devices to situate a robot in its spatial and operational contexts: its location relative to where it must go and what it must avoid and its operating parameters such as temperature and moisture.

UAV (Unmanned Aerial Vehicle)
Remotely piloted aerial platform, commonly referred to as a "drone," deployed by the U.S. armed forces for both reconnaissance and the delivery of ordnance.

FURTHER READING

Brooks, Rodney. *Flesh and Machines: How Robots Will Change Us.* Cambridge, MA: MIT Press, 2002.

Brynjolfsson, Erik, and Andrew McAfee. *The Second Machine Age: Work, Progress, and Prosperity in a Time of Brilliant Technologies.* New York: Norton, 2014.

Eggers, Dave. *The Circle.* New York: Knopf, 2013.

Kurzweil, Ray. *The Singularity Is Near: When Humans Transcend Biology.* New York: Viking, 2005.

Lanier, Jaron. *You Are Not a Gadget: A Manifesto.* New York: Knopf, 2010.

Markoff, John. *Machines of Loving Grace: The Quest for Common Ground Between Humans and Robots,* New York: Ecco, 2015.

Nourbakhsh, Illah Reza. *Robot Futures.* Cambridge, MA: MIT Press, 2013.

Reeves, Byron, and Clifford Nass. *The Media Equation: How People Treat Computers, Television, and New Media Like People and Places.* New York: Cambridge University Press, 1996.

Singer, P. W. *Wired for War: The Robotics Revolution and Conflict in the 21st Century.* New York: Penguin Books, 2009.

INDEX

JOHN JORDAN is a technology analyst and Clinical Professor of Supply Chain and Information Systems in Smeal College of Business at Penn State University.